Science as Psychology

SENSE-MAKING AND IDENTITY
IN SCIENCE PRACTICE

Lisa M. Osbeck

University of West Georgia

Nancy J. Nersessian

Georgia Institute of Technology

Kareen R. Malone

University of West Georgia

Wendy C. Newstetter

Georgia Institute of Technology

CAMBRIDGE
UNIVERSITY PRESS

CAMBRIDGE
UNIVERSITY PRESS

32 Avenue of the Americas, New York NY 10013-2473, USA

Cambridge University Press is part of the University of Cambridge.

It furthers the University's mission by disseminating knowledge in the pursuit of education, learning, and research at the highest international levels of excellence.

www.cambridge.org
Information on this title: www.cambridge.org/9780521708418

First published 2011
First paperback edition 2013

A catalog record for this publication is available from the British Library.

Library of Congress Cataloging in Publication data
Science as psychology : sense-making and identity in science practice / Lisa M. Osbeck . . . [et al.].
    p. cm.
Includes bibliographical references and index.
ISBN 978-0-521-88207-1 (hardback)
1. Engineering – Research. 2. Problem-solving – Social aspects. I. Osbeck, Lisa M. 1962–
TA160.S35 2010
620.0072–dc22   2010028652

ISBN 978-0-521-88207-1 Hardback
ISBN 978-0-521-70841-8 Paperback

CONTENTS

Acknowledgments *page* vii

1. Introduction: Science and Persons   1
2. Methods of Study   31
3. The Problem-Solving Person   52
4. The Feeling Person   92
5. The Positioning Person   120
6. The Person Negotiating Cultural Identities   157
   Part I: The Person Enacting Race   162
   Part II: The Person Enacting Gender   179
7. The Learning Person   195
8. Epilogue: Science as Psychology: A Tacit Tradition and Its
   Implications   219

References   249
Index   271

ACKNOWLEDGMENTS

We gratefully acknowledge the significant support we have received from the following sources: National Science Foundation, ROLE and REESE Programs of the Division of Research on Learning: REC0106733, DRL0411825, and DRL097394084 (Nersessian, PI; Newstetter, co-PI; Malone, Associated Faculty; Osbeck, Research Scientist); National Endowment for the Humanities (Nersessian); Radcliffe Institute for Advanced Study (Nersessian); Spencer Foundation (Malone); and Georgia Institute of Technology, Faculty Development Grant (Osbeck). The opinions expressed in this book are our own and not those of the funding agencies.

Some portions of this book draw from previously published material: Nersessian, N. J. (2009). How do engineering scientists think? Model-based simulation in biomedical engineering laboratories. *Topics in Cognitive Science*, Vol. 1, No. 4, 730–757; Osbeck, L. (2009). Transformations in cognitive science: Implications and issues posed. *Journal of Theoretical and Philosophical Psychology*, Vol. 29, No. 1, 16–33; Osbeck, L. & Nersessian, N. (2010). Forms of positioning in interdisciplinary science practice and their epistemic effects. *Journal for the Theory of Social Behaviour*, Vol. 40, No. 2, 136–161; and Osbeck, L. & Nersessian, N. (2010). Affective Problem Solving: Emotion in Research Practice. *Mind and Society*, Vol. 9, No. 2. We thank the various presses for their permission to use the material.

We thank the directors of the laboratories we studied for opening their labs to us and for being willing to discuss their research and ours during the course of the investigation. We further thank the lab members for welcoming us into their environment and generously participating in numerous interviews and clarifying discussions. We thank Dr. Gilda Barabino for invaluable collaboration on the gender and race research with Kareen Malone. Finally, we thank the members of our extended research team who helped make this book possible, especially Robin Atkinson, Susan Bernard, Sheila Burgess, Sanjay Chandrasekharan, Jim Davies, Nonah Elliston, Michelle Grandin, Ellie Harmon, Shannon Kelly, Elke Kurz-Milcke, Christopher Patton, and Paul Tullis.

1

Introduction: Science and Persons

George Kelly used "scientist" as a metaphor for "person" to emphasize that understanding science practice – how scientists think and what they do – enables us to understand human nature more profoundly. In turn, understanding human nature invites critical appraisal of our notions of science as an activity of persons:

> Psychologists are likely to be very much in earnest about making their discipline into a science. (Unfortunately, not many are as concerned as they might be about making science into something.) . . . But what would happen if one were to envision all human endeavor in those same terms the psychologists have found so illuminating in explaining themselves to their students? And indeed, might it not be that in doing so one would see the course of individual life, as well as human progress over the centuries, in clearer perspective? Scientists are men, and while it does not follow that men are scientists, it is quite appropriate to ask if it is not their human character that makes scientists what they are. This leads us to the question of how that human character can be better construed so as to account for scientists, and whether our construction can still explain as well the accomplishments that fall far short of what we, at this transient moment in our history, think good science is (George Kelly, unpublished manuscript quoted in Bannister & Mair, 1968, pp. 2–3).

In a similar spirit, but with differing concepts and method, this book is written with the conviction that science practice provides fertile yet undercultivated ground for psychological theorizing. The central question with which psychology historically has wrestled – how best to characterize the integration of the bodily, intentional, environmental, social, and cultural dimensions of human life in day-to-day functioning and long-range achievement – is spotlighted and amplified in the microcosm of the science laboratory. Thus the laboratory is psychologically important not merely because it allows for isolation and control of variables, as is psychology's typical view of the laboratory's investigatory benefits. Rather, the laboratory is important because, within the culture of even a *single* laboratory, cognitive, social, affective, material, and other dimensions of human activity are richly and importantly interlaced in a "mangle of practice," as Pickering (1995) has called it.

In *Epistemic Cultures*, Knorr Cetina proposed a view of the laboratory as "an 'enhanced' environment that 'improves upon' natural orders in relation to social orders" (1999, p. 26). Sciences display "the *smear* of technical, social, symbolic dimensions of intricate expert systems" (Knorr Cetina, 1999, p. 3, emphasis added). Although the focus of science studies is on understanding science and scientific knowledge for its own sake, we are suggesting that natural science is also an especially informative locus of human activity, the inherent complexity of which offers inroads for understanding human nature and functioning more profoundly. This book reflects our view of science or rather of scien*tists* – the activities and articulations of persons working as scientists in situ – as a relatively untapped source for generating and honing ideas we categorize as "psychological." As we shall discuss, this is a new reading of the "person-as-scientist" metaphor forwarded by George Kelly in the 1950s.

In this effort we describe several specific psychological dimensions of science practice through our analysis of the accounts and activities of working scientists. The scientists we analyze are biomedical engineers collaborating within well-regarded laboratories on the campus of a major American research university.[1] Despite the long-standing demarcation of "basic" or "pure" science from applied or problem-focused professions such as engineering, the laboratories we study are innovation communities engaging in cutting-edge research that transcends and blurs any boundaries between basic and applied science. The focus of our analysis is a set of interviews with researchers possessing varying levels of expertise, from students entering laboratories for the first time to the principal investigators who have established the laboratories under study. The work here builds both on extensive analysis already undertaken and on research that is ongoing, the details of which are given in Chapter 2.[2]

Some background discussion is necessary to clarify how our analytic efforts offer a departure from traditional conceptions of the relation between

---

[1] Details about the scientists and the form of science practiced are provided in Chapters 2 and 3.

[2] See, for example, Nersessian (2006); Nersessian and Chandrasekharan (2009); Nersessian, Newstetter et al. (2003); and Osbeck and Nersessian (2006).

psychology and natural science. We begin by drawing a distinction between psychology as science and science as psychology and then address the problems to which our analysis is directed and the ways in which it is intended as a contribution. We offer an outline of the book as a whole at the end of this chapter.

PSYCHOLOGY AS SCIENCE VS. SCIENCE AS PSYCHOLOGY

As a starting point, it is important to note that "psychology as science" has functioned as something between a slogan and mandate for the academic and professional community of psychologists for more than a century and a quarter. The assumption it crystallizes is that psychological knowledge is most trustworthy and prestigious when psychologists pattern their methods of inquiry after those presumed to be the bedrock of natural science. Thus researchers and practitioners hold each other accountable to the normative framework of "scientific psychology," with controlled experimentation the beau ideal; disciplinary hierarchies follow. In turn, the mandate has sparked assorted forms of rebellion. Reactions to the methodological strictures and perceived philosophical vacuity accompanying a too rigid psychological scientism include the mid-20th-century eruption of Third Force psychologies and contemporary alternatives devoted to critical and hermeneutic psychologies. Indeed, nothing has been as polarizing to the discipline of psychology as its ideas *about* natural science, whether in relation to psychology's potential for achieving a full measure of objectivity or to the ontological status of the objects of both human and natural science. Psychologists operate within a field defined by poles representing, on the one hand, a largely unexamined emulation of science and, on the other, the marginalized position of critic. Genuine dialogue between those positioned at either end occurs most frequently around methodologies, over the appropriate combinatory applications of qualitative and quantitative research designs in relation to a specific problem (e.g., Tashakkori & Teddlie, 2002). Despite the advantages offered by mixed method designs, psychologists should not conflate the mixing of methods with meaningful conceptual integration.

Psychology needs new questions concerning its *range of possible relations to science* as a way to revitalize the inquiry. A step in this direction is to bypass or bracket the question of whether psychology can or does have proper scientific standing. One means of changing strategies is to focus psychological attention on the quite fascinating human practices that constitute science itself, thereby helping inform the persistent question about what science is in the first place. If we at least temporarily suspend the debate over psychology and science and leave behind preconceptions about the latter, psychologists have much to mine from the everyday business of scientists in practice: their complex learning trajectories, their myriad creative efforts to achieve coherence and develop new concepts, the entanglement of their sense-making with emotional engagement and values, and the shifting, intricate identity formations negotiated in relationship with one another and with the material culture of their practice. Thus in this text we engage natural science not to borrow its methods but to bring psychological questions to its practice. The central task of our analysis is to consider what might be mined theoretically by turning the psychology-as-science mantra on its head – by imagining *science as psychology*.

Under the direction of principal investigators Nersessian and Newstetter, we have formed an evolving, interdisciplinary team of ethnographers to describe and understand the learning, reasoning, and problem-solving practices of both the novice and expert researchers in two biomedical engineering laboratories, details of which are provided in Chapter 2. We draw both from ethnographic observations and analysis of a large set of interviews conducted with research scientists in the two laboratories. The interviews concern the nature of the scientists' work, the problems they are working on, the sources and progression of their ideas, their learning and social experiences in the laboratory, transformations in their identity through their encounters with persons and objects in the laboratory, and their aspirations and plans for the future.

Here one might legitimately question why the practice of *science* should be targeted for psychological analysis rather than other forms of professional, skilled, or nonskilled practice. Part of the answer has to do with psychology's emulation of science and what this reveals about the discipline's self-representation. Psychology's natural science aspirations spring from conviction of the special standing and ultimate authority of science, a conviction that is certainly not limited to the discipline of psychology. Philosophically, it is tied to the ideal of science as a value-free or value-neutral enterprise, on which there is a vast literature with roots traceable to Bacon and Galileo (Lacey, 1999). Because of this ideal, science is enveloped in mystique that more than one author has branded *mythical*, even religious in its overtones: Kitcher (1993) speaks of the "legend" view of science, Mitroff (1974) acknowledges the "storybook" view, and Mahoney calls the scientist the "high priest of knowledge" (2004, p. 3). The reason we target science practice in particular has to do with the relation science seems to bear to our highest human ideals: what we value in ourselves and see as the farthest reaches of our intellectual power.

There have indeed been prominent psychological efforts to embrace scientific thinking as prototypical of human rationality, and rationality as the quintessential human attribute. Among the arguments that Greg Feist's (2006a) pioneering *Psychology of Science and the Origins of the Creative Mind* offered as a rationale for a new subdiscipline organized around psychological dimensions of science is that science and scientific thinking constitute "prototypes of human thought and understanding" (p. 3). Note, however, that there are two quite divergent connotations of "prototype": one implying an idealized form of something, the other a typical representative of a class. As we shall discuss in our final chapter, two quite varied accounts of scientific thinking correspond to these two connotations. In turn, the different accounts of scientific cognition give rise to two different representations (depictions/perspectives) of scientists and by extension the human characteristics that scientists represent.

There have been scores of efforts to cast doubt on the possibility of a value-free enterprise of any kind and of a value-free science specifically, even if these efforts have yet to have an impact on the business as usual of psychology practice. Our aim in this work is not to dismantle the rational core or privileged position of science – far from it. Yet if our analysis of one particular context of scientific practice might provide some insight into how people "do" rationality in research laboratories – that is, in the settings upheld as demanding the purest of rational practices – we might be better poised to offer some comment on the nature of rational operations more generally. Note also that our intent is not to reinvent the wheel crafted by the extensive body of literature in philosophy of science, cognitive science, and related disciplines that demonstrates the embodied, embedded nature of rational functioning. Rather, we aim to clarify how some of the rich insights of recent science studies might have relevance to the psychological community and its questions and, at the same time, supplement existing lines of theorizing with our own analysis of a particular research culture.

Problems Targeted

We organize our analysis of interview content and observational data to address two related and overlapping problems relevant to the study of science and the discipline of psychology. The first is what we identify here as the *integration problem*: the challenge of characterizing adequately the fluid entanglements of cultural, cognitive, affective, and other dimensions as they coordinate in all forms of human activity and particularly in those for which targeted problem solving is a principal aim (see Nersessian, 2005). As Lave noted, the very notion of "problem solving" historically invites frameworks (and, until recently, methods) that exclude social, cultural, and emotional considerations: "Today, for example, it is likely to be assumed that if an ongoing activity consists of problem solving – 'individual, rational, cognitive' – it is not necessary to address the possibilities that it is culturally and socially structured, primarily expressive of feelings, or part of socially contextualized experience in ways that require theorizing, empirical description, or analysis" (1988, p. 7).

The second problem we target is the unsatisfactory status of contemporary general psychology, which instead of serving as a foundation for subdisciplines through its questions and analytic strategies has devolved into a grab-bag of specialized lines of research. We call this the *problem of grab-bag general psychology*.[3] Our assessment is a commentary on the status with which general questions about human functioning and integrated efforts to address them tend to get taken up by the academic community, outside of the discipline of psychology but more egregiously within it. Instead of upholding general psychology as the axis of the most thorny and intriguing challenges for the discipline of psychology, the convention in psychology departments is to limit it to an introductory-status course, one eagerly farmed out to graduate students for the purposes of cutting their teaching teeth. Students are left with the impression that psychology is a collection of research findings conveniently parsed into traditional categories around which academic careers have been organized. Indeed this is not an inaccurate assessment of the state of the field. Yet general psychology as a *theoretical pursuit* can be much more and deserves to be so. It is among our goals to contribute to its revitalization.

To summarize, central to the analysis and the organizational scheme employed for its presentation here are two related claims:

1. Our study of scientists offers an opportunity to present carefully *integrated* accounts of cultural, cognitive, and affective human dimensions of human activity. We thus align our efforts with those of theorists both recent (Papadopoulos, 2008) and historical (Vygotsky, 1978) who argue that we cannot understand volitional, cognitive, and emotional phenomena as isolated systems but rather must examine their concrete and everyday realization in human action and community.

---

[3] Note that the naming of this problem is not intended as a criticism of the *Review of General Psychology* (the APA Division 1 journal), which explicitly endorses the importance of cross-cutting research in its editorial aims.

2. Thus the study of science *as practiced* has the potential to inform questions concerning human functioning and experience more broadly, that is, to inform the project of a general psychology.

We begin with a brief review of how we intend this text to address both the *integration problem* and the *problem of grab-bag general psychology* and then present the outline of the book as a whole and the content of its chapters. Our hope is that the analysis offered here will not only serve as a contribution to science studies, specifically to the burgeoning field of psychology of science, but more directly as a contribution to a freshly envisioned general psychology.

SCIENCE STUDIES AND THE PSYCHOLOGY OF SCIENCE: THE INTEGRATION PROBLEM

An infrequently recognized source of the conflict surrounding the question of psychology's scientific status is that although "science" does not lend itself to facile definition or understanding, psychologists have attempted to impose plainness on it by equating science with a rigidly codified method (equated with "The Scientific Method"). Thus the question of whether psychology is a science, or what kind of psychology is scientific and what is not, is addressed by reference to what is methodologically sanctioned. But scholarly views of science are quite complex. Academic dividing lines are at least chalked around the question of how science is to be understood, both within and between disciplines devoted to the study of science. At one end are accounts emphasizing a set of core similarities and an essential inner logic common to all forms of science, which form a foundation of enduring rational structures to which the norms of particular sciences ultimately appeal (e.g., Carnap, 1935; Hempel, 1952). More recent emphasis has been placed on the grounding of these structures in the sound operations of mechanisms detailed by cognitive science – this is the focus of cognitive studies of science.[4] At the other end, rational and cognitive descriptions of science have long faced competition from accounts of science as a fundamentally social and even political system – from the argument that the logic of science itself reflects deeply ingrained human habits and negotiated rules to the assertion that there are broad institutional and economic

---

[4] For example, Carruthers, Stich, and Siegal (2002 [edited volume]); De Mey (1982); Giere (1988, 1992); Gooding (1985); Gorman et al. (2005); Klahr (2000); Nersessian (1984, 2008); and Tweney, Doherty, and Mynatt (1981).

forces that inflect the forms of representation that influence how methodology is understood and our material reality is interpreted. Such forces may
even be seen as determining the questions to be asked and the explanations
proffered.[5]

Rational–Social Divides

That, historically, socio cultural and rational-cognitive accounts of science
have been at epistemological loggerheads is obvious in the use of the term
"science wars" to describe their relationships. Writing in the early 1980s
and characterizing 20th-century trends, Marc De Mey remarked on the
fragmentation of science studies resulting from the competing interests of
philosophy and social sciences:

> Philosophy of science claims a special position because of the special
> nature it attributes to scientific knowledge as superior knowledge, this
> special nature being essential for justifying its existence as a separate
> discipline. Sciences of science such as the psychology and sociology of
> science, on the contrary, appear to consist of the application of empir-
> ical disciplines not developed with science as their primary object of
> study, and their products, though sometimes very penetrating and highly
> interesting, seem unrelated to each other and leave us with a picture of
> the science of science as a rather fragmented endeavor (De Mey, 1982,
> p. xv).

The picture of something rather fragmented remains when surveying the
development of science studies since the time of De Mey's writing. A kind
of grafting of cognitive and neuroscience research onto the rational reconstructions provided by philosophy of science has occurred over the past
three decades. Although controlled and naturalistic observation have added
immeasurably to philosophical accounts of model use, analogy, visualization, and metaphor in scientific reasoning, theory formation, hypothesis
testing, and discovery, the vast majority of cognitive studies of science have
proceeded in relative isolation from social and cultural studies of science,
robust and internationally poised as they also may be. More recently, Helen
Longino (2002) and Nancy Nersessian (2005) separately have pointed to the
implicit acceptance of a rational-social dichotomy in philosophy of science
and science studies, prodded by critical reactions to the reactionary estab-
lishment view of science as a rational, rule-bound, progressive enterprise

---

[5] For example, see Feyerabend (1975); Latour and Woolgar (1986); Levins and Lewontin
(1985); and Pickering (1984).

driven by the highest cognitive achievements. The fallout is that science tends to be interpreted as *either* a cognitively *or* a socially powered phenomenon, and accounts of science typically are rational-cognitive *or* social accounts (i.e., the categories of rational and social are sharply distinguished). Yet the tendency to dichotomize social and cognitive accounts of science invited by the academic divide is both conceptually problematic and at odds with the complexities of science underscored by historians and ethnographers of science.[6]

Integrative Efforts

Several key lines of integrative effort serve as important counters to the trend of disengagement of social and cognitive accounts. The entire project of cognitive anthropology is a ready example,[7] as is ethnomethodology as the study of reasoning in situated contexts of practical activity.[8] In large part borrowing from these frameworks, cognitive scientists increasingly have been investigating reasoning and problem solving in naturalized settings in which the local features of the context, including social roles and cultural artifacts, are theorized as essential to the cognitive tasks at hand. In the case of extended mind theory, those local features are considered part of the cognitive process itself.[9]

Since the 1970s, feminist scholars have undertaken many efforts to articulate more integrated formulations of science. Often derived from the experience of practicing scientists or stemming from questions concerning the nature of objectivity, such accounts forcefully argue for less individualistic notions of knowledge generation, a better understanding of the role of affect in relation to knowledge, and a keener appreciation for the role of community and intersubjective relationships in generating knowledge and criteria for judgment.[10] Although often stereotyped as purely interested in the social dimension (qua gender), such accounts frequently attempt to construct a

---

[6] Nersessian (2005), who has engaged in extensive historical and ethnographic study of several sciences, has forthrightly declared the divide artificial: "Producing scientific knowledge requires the kind of sophisticated cognition that only rich social, cultural, and material environments can enable" (p. 18). She has devoted her recent research efforts to developing ways to bypass or dissolve the traditional dividing lines to theorize culture and cognition as aspects of the same infinitely complex system (Nersessian, 2005, 2006).

[7] For example, see history by D'Andrade (1995) and Shore (1996).

[8] For example, see Garfinkel (1967); Hutchins (1995a, 1995b); Lave (1998); and Lynch (1993).

[9] See Clancey (1997); Clark (2003); Suchman (2007); and Wilson (2004).

[10] See Haraway (1991); Longino (2002); and Nelson (1990).

broader conception of scientific cognition through questions such as "What sort of person is the scientist?"

Omissions and Oversights

Yet vitally important things continue to fall out of cognitive and social accounts, even accounts that manage to cross the traditional social and cultural divide. Three omissions are especially noteworthy:

1. The material grounding of science practice – not only the neural and other bodily processes of scientists but also constraints imposed by the nature of the objects and artifacts to which their practices are directed – remains challenging to incorporate smoothly with accounts of linguistic, social, and normative dimensions of practice.[11] Thus some recent contributions to cognitive science construe culture, material, and social environments as aspects of a single system of processes (e.g., Hutchins, 1995b; Nersessian, 2005, 2006). Collectively these have been called "environmentalist" approaches in cognitive science (e.g., Nersessian, 2005; Rowlands, 1999).

2. Neither thoroughgoing social, cognitive, or even blended social and cognitive accounts are easily able to account for the contribution of the *particularity* of the scientist, that "something else" of embodied and storied living persons that for want of better options we resort to conventional psychological categories such as affect, motivation, personality, and subjectivity, but that appear to be an indispensable feature of at least science in process. That is, science is shot through with what is irreducibly a matter of personal *style* and, in addition to *norms*, science includes a heavy dose of *value*, and more provocatively, "*desire*." The contribution of this additional dimension – this "something else," this particularity – to the practices and processes that constitute science is more or less ignored in the mainstream of science studies, at least outside of biographical analyses of scientists.

3. The influence of the subjective or personal dimension is acknowledged principally as a source of error or complication. The "personal equation" in astronomy, for example, referred to individual differences in reported stellar transit times among astronomers using the same instrument, which muddled the effort to provide precise estimations of distance (see Duncombe, 1945). Simon Schaffer has referred

---

[11] See Daston and Galison (2007); Galison (1997); Maquet (1993); and Pickering (1995).

to this as "the personality problem" (1988). In psychology's early history, Titchener's "stimulus error" referred to the tendency to slip from the "psychological" point of view (equated with the "scientific") to another point of view contaminated by personal interpretation or "meaning" (Titchener, 1912, p. 488). Even scientists themselves tend to evoke the role of personal or subjective factors only in relation to failure or deviation from protocol (Gilbert & Mulkey, 1984). What is missing is a general (theoretical) account of the ways in which this dimension contributes to science practice in both the discovery and justificatory phases; that is, in ways that enhance science, not merely detract from its purity and precision.

Science and Personality

The traditional view is that the "something else" is the psychologist's responsibility, and indeed a few psychologists have conducted assessments of the personality attributes of scientists; others have examined the nature of scientific creativity. For example, personality traits associated with science as a choice of profession form the focus of Bernice Eiduson's (1962) *Scientists: Their Psychological World*. Eiduson employed clinical interviewing and projective testing to analyze personality features of scientists as part of a larger project comparing the analysis of scientists to her earlier analysis of artists, writers, and musicians. Greg Feist, well known for his contributions to both personality psychology and the psychology of science, reviewed evidence suggesting that certain tendencies of disposition and temperament "function to lower thresholds for interest, talent, and achievement in science" that "make becoming a scientist more likely and influence both the kind and quality of scientist one becomes" (2006a, p. 115).

Feist's review efforts include a meta-analysis of twenty-six published studies in which he compared personality traits of scientists or science students with those of persons not oriented toward science (Feist & Gorman, 1998). He grouped emerging tendencies into cognitive, social, and motivational categories and identified higher conscientiousness and higher openness to experience as cognitive traits common to the scientific personality, with higher dominance and assertiveness (possibly even hostility) and lower social interest and affiliative tendencies as social traits, and achievement orientation and ambition as motivational traits characteristic of the science-oriented group. He further analyzed personality tendencies associated with particular science specialties, oriented around a primary distinction between people-orientation and thing-orientation that stemmed originally from classical vocational interest literature (Holland, 1973; Prediger, 1982). Feist also made note of recently identified links between high scores on measures of autistic spectrum features and spectrum conditions and an ability and interest in science, especially engineering, mathematics, and physical science (e.g., Baron-Cohen et al., 1998). Elsewhere (1998, p. 96), Feist noted that self-image and "demographic forces" including birth order, religious affiliation, and immigrant status also appear to influence scientific interest.

Feist recognized, as have others, that interest in science and achievement in science are very different matters, and thus his meta-analysis also included a comparison of traits of scientists whose contributions were judged to be particularly eminent and innovative. He also interviewed more than one hundred members of the National Academy of Sciences in addition to administering personality questionnaires and having research assistants conduct blind evaluations of audiotapes. He concluded that many of the traits that appear to be associated with scientific interest are more pronounced in those who demonstrate particular scientific creativity. However, he also identified a positive association between hostility and arrogance and particular scientific achievement. Perhaps most interestingly, in contrast to the image of the plodding, methodical, rule-following scientist that sometimes emerges in cultural critiques of scientism, he wrote,

> Conscientiousness is not a defining trait of the most distinguished scientists – at least when compared to the average scientist. Indeed, conscientiousness and creativity, although not polar opposites, are somewhat opposing forces of personality. There is no doubt some truth to the argument that some degree of conscientiousness is required to turn creative potential into actual creative products – the discipline and stick-to-itiveness to see the task to completion. But the traits of careful, cautious, conventional, disciplined, orderly, persevering, reliable, and self-controlled are less evident in the most creative of scientists (Feist, 2006a, p. 123).

Dean Keith Simonton (1988, 2004) has similarly analyzed the personality traits of especially creative scientists. These efforts are related to studies of personality and giftedness more generally (Ferriman, Lubinski, & Benbow, 2009; Winner, 2000).

Interesting and informative as analyses of this kind prove to be, the application of trait-based personality psychology to science studies upholds an implicit assumption that the scientist is a distinct "type" of person. Notwithstanding the folk impression and repeated empirical finding that certain constellations of dispositional tendency accompany an interest in science and that others might predict the ability to make an original contribution to science, personality psychology has not told us everything that is interesting about the "something else" in science practice. That is, identification of personality traits more prevalent in scientists does not allow us to consider the intricate means by which the emotional life and storied particularity of the scientist contribute moment by moment to the web of cognitive and social practices that constitute science.

Of course, historians of science have for many years analyzed on a case-by-case basis the interplay of the scientist's personal history, emotional life, and particular cognitive style in the insights eventually attained. Intellectual biography is the most obvious genre in which this interplay is illustrated, such as in Isaacson's recent and acclaimed biography of Einstein (Isaacson, 2007). Yet even when the focus is on the models used and generated by a given scientist of note, the peculiar contribution of the embodied and historied person is everywhere entwined with the cognitive analysis.[12]

Toward Enhanced Integration

A disciplinary effort that shows early potential for integrating accounts of science is the formation of a new international group of scholars that has organized recently under the banner of "psychology of science." Although psychological studies of science can be traced to the 1870s (see Feist, 2006b), it was not until 2006 that the International Society for the Psychology of Science and Technology was formed through the organizational efforts of Greg Feist, Michael Gorman, and Sofia Liberman. This effort is promising not only because "psychology of science" remains an unfamiliar term and territory but also because the society's formation offers opportunities for dialogue between traditionally divergent approaches to science. Theoretical diversity among society meeting attendees and the editorial board of the society's new journal is facilitating the interactive nature of exchanges. The full range of questions and methods is yet to be determined.

As Greg Feist has pointed out, psychology has lagged quite severely behind other fields of science studies such as sociology of science, philosophy of science, and history of science (Feist, 2006a). We might note that American psychology in particular has as yet made much less visible contributions to fields of science studies than have other fields. Although cognitive

---

[12] See Gooding (1985); Nersessian (2002); and Tweney (1992).

studies of science are well developed through cognitive science and cognitive psychology, there has been little effort to integrate these with social studies of science or at least to encourage the peaceful coexistence of cognitive and social studies under an awning of psychological studies of science (e.g., see Longino, 2002; Nersessian, 2005). Moreover, there have been infrequent efforts to examine contributions that psychologists are uniquely poised to offer, such as the analysis of personality, emotion, motivation, identity, and personal style as contributions to science process. Drawing a contrast with the sociology of science, Michael Mahoney claimed that "the psychology of science can hardly claim a skeleton let alone some flesh. . . . Two words summarize our knowledge: nominal and sketchy" (Mahoney, 2004, pp. 28–30). The term "psychology of science" remains an unfamiliar one despite several attempts to establish it as an interest area in years past.[13] There are notable exceptions, but by and large explicit psychological interest in science remains relatively obscure.

However, Feist has argued that many distinctly psychological contributions might be legitimately claimed as "implicit" contributions to science studies; that is, much research in neuroscience, cognition (particularly thinking, problem solving, insight, and creativity), social psychology, development, and even personality psychology brings important findings to the understanding of science and scientists. The subfield is in the very early stages of organization and planning, with international momentum building to develop the psychology of science into a recognized subfield of both psychology and science studies.

One factor that might limit the visibility of psychology's contribution to science studies is methodological. Increasingly, cognitive scientists study cognition in naturalized settings, or cognition "in the wild," as Hutchins (1995b) termed it. By contrast, psychologists have long defined their own disciplinary distinctiveness in terms of a methodology in which the isolation and control of variables are mandatory. Although arguably of late they have displayed more tolerance for qualitative methods, psychologists remain inclined to regard qualitative data suspiciously or at least as inferior to data that can be controlled and from which results might be justifiably generalized. Thus the psychology of science, though offering great potential for integrated accounts of science, remains dominated by traditional psychological method and a corresponding theoretical focus on individual-level processes. Psychology at large has not keep pace with innovations in cognitive science, and its attempts at

---

[13] For an historical review, see Feist (2006a, pp. 23–24) and Feist & Gorman (1998).

integration have been methodological (i.e., through methods deemed scientific) rather than conceptual.

Exciting new developments in cognitive science aim purposefully to provide carefully integrated accounts of human reasoning and problem solving that give body, cognitive function, and social and material environments shared responsibility in the production and dissemination of knowledge; these accounts generate new concepts for characterizing relations between these domains (e.g., Clark, 2003; Menary, 2007). As Rowlands (1999) pointed out, an ontological claim supporting what he termed environmentalism in cognitive science concerns the location of cognitive processes (i.e., not exclusively inside the skin); an accompanying epistemological claim is that such processes cannot be understood by focusing on the area under the hood in isolation. Thus the unit of analysis is the intelligent act or practice as performed in real-world settings, in local practice communities: "human cognition in its natural habitat" (Hutchins, 1995a, p. xiii). The practices in question, sometimes called "cognitive practices," include applications of memory, problem solving, and model construction and communication in real-world contexts of both routine performance and innovation. The increased use of naturalized settings and ethnographic methods by cognitive scientists affords further opportunities for a hammering out of social-cognitive crossover language useful for descriptive and theoretical purposes. As one example, Nersessian and colleagues have previously articulated the concept of "cognitive partnering" to capture a notion of cooperative participation in learning and reasoning practices situated in biomedical engineering labs that is essential to conceptual change and innovation.[14]

Nersessian, Newstetter, and colleagues have been instrumental in bringing integrated approaches into the study of research science laboratories, in particular of engineering scientists in two cutting-edge biomedical engineering research laboratories located on the campus of a major research university. Biomedical engineering (BME) is a relative newcomer to science; it is a hybrid specialty that blurs the disciplinary boundaries of biology and engineering and the pure and applied traditions of science as traditionally distinguished. The particular laboratories we study in this text aim to develop new understanding of biological phenomena, methods, and applications at the frontier of BME research. As innovation communities in this sense, the laboratories afford exceptional opportunities to observe and analyze scientific problem solving and creativity in action. As a community

---

[14] See for example, Nersessian, Kurz-Milcke, et al. (2003) and Osbeck and Nersessian (2006).

or culture, each lab also offers rich opportunities to theorize social inter-actions and the ways in which these interactions fuse with and facilitate cognitive practices. We have been studying researchers at varying levels of expertise, from novice undergraduate students obtaining their inaugural laboratory experience to the principal investigators (PIs) of large-scale, fed-erally funded, multiyear research efforts. Under the direction of Nersessian and Newstetter, we have spent many hours observing these researchers' work and interviewing them about their experience and many more hours thinking about how best to analyze the material. We have conducted our investigation as an evolving team with diverse disciplinary backgrounds and commitments. The data collected in that research provide the primary con-tent of our analyses in this book. Specific details about the laboratories and researchers as well as our investigatory procedures are provided in Chapter 2 (Methods).

Thus far we have focused on the problem of integration in science stud-ies, principally the integration of cognitive and socio-cultural accounts of science. We have also considered what might fall through the cracks of these efforts and have highlighted the potential for more thorough integration through recent developments in cognitive science, including analyses of problem solving in real-world contexts of scientific practice. The question we now take up is *how* a sufficiently integrated, adequate account of science might be informative beyond the laboratory – specifically, how it might contribute to psychology.

The Problem of Grab-Bag General Psychology

Like science, "psychology" covers an elaborate yet ambiguous landscape: "The material is too great in mass and too varied in style to fit into existing pigeon-holes, and the cabinets of science break of their own dead weight" (Dewey, 1896, p. 357). Therefore it is important to be specific about the psychology to which we aspire.

The psychology that emerges from our study of BME laboratories is a *general* psychology, of late a foreign-sounding and frequently disparaged term. Too often "general" psychology is interpreted either as lacking in detail or as a loose conglomeration of divergent research strands with few struc-turing interconnections. Despite frequent calls for unification and awards for cross-cutting efforts offered by the Society for General Psychology (APA Division 1), unification is frequently imagined through a particular version of methodological rigor, in keeping with perceived natural science rigor, as outlined earlier. Robert Sternberg, among the most visible critics of psychology's fragmentation, has derided the discipline's segregated theoretical structure, by which organizing theories are weeded out through increasingly specialized and narrowly focused experimentation (Sternberg, 2001). Theoretical unification is often yoked with a particular vision of science, and it is often suggested that it comes at the price of a methodological hegemony that stifles the range of questions that can be asked and legitimately claimed as psychological.[15]

We take inspiration for our conception of a more robust general psychology from the neglected tradition of inquiry relevant to a foundational, holistic, and integrated understanding of persons (e.g., James, 1890; Stern, 1906, 1917). The intent of early general psychologies was not only to survey the field comprehensively but also to identify *units of analysis* appropriate to the complexities of human functioning and experience, which were embedded in overarching philosophical commitments recognized as inescapable.

Example: Stern's *Unitas Multiplex*

For its illustrative potential we briefly turn our attention to the philosophically informed tradition of critical personalism associated with the German psychologist William Stern. Stern's work has been long overlooked but has enjoyed renewed attention principally because of the scholarly work of personality psychologist James Lamiell (2003, 2009). In Lamiell's translation, Stern's *unitas multiplex* refers to "the multiplicity of characteristics related to the unity of the individual" and the "immanent coordination of parts within the whole" toward "a real, goal-directed self-activity" (Lamiell, 2010, p. 113). Within this general framework, four levels or "layers" are distinguished – phenomena, acts, dispositions, and subject – not as quantitatively distinct but as "different logical categories," each involving a differential "field of projection" of the person, the primary metaphysical reality for Stern (1906, p. 11). Thus, intriguingly, Stern considered the concept of *unitas multiplex* to hold "the key to science in general as well as to the science of the human being" (p. 6):

> Within each act, a multiplicity of phenomenal experiences is formed into a unity; in each disposition, a multiplicity of ever-recurring acts is grounded within a domain of possibilities; in each subject a multiplicity of dispositions comprises a unified individual. None of these levels may be reduced to one of the others in the interest of simplifying the psychological system. Each is necessary in its categorically special nature. . . . On

[15] See Kirschner (2006); Koch (1993); and Robinson (2007).

the other hand, nor may the levels be reified as separable parts of the psychological individual, because there are indissoluble causal and goal orientation relationships that bind the layers to each other and, above all, to the ultimate, the unified individual. This alone is – as person – what is objectively real (pp. 11–12).

Representations, Processes, and Stimulus–Response (S-R) Connections vs. Acts, Actions, Activities, and Practices

As Stern's account of person begins to suggest, a fully integrated account of personhood requires more than a sense of the coordinated working of variegated human capacities. A yet more byzantine challenge is to establish units of analysis appropriate to the complexities of *mind-world* relations, for which he offered the additional principles of convergence and introception. Importantly, recent efforts to establish the person as metaphysically primary are grounded in the linguistic analysis of the category of personhood; that is, the conception of person is tied to language-using practices and forms of rationality that they engender that distinguish our species from any other.[16] Because language is inherently social, the focus necessarily expands beyond that of the individual, rational agent into the social, normative world in which the individual is a participant. To date the mainstream of psychology has failed to meet the challenge of adequately theorizing mind-world complexities. For a half-century, cognitive psychology has provided the framework for understanding psychological processes at large, within which the focus of psychological analysis has been the cognitive state or process, typically understood as an individual-level event.

Indeed, for many years the default position of contemporary psychology has been uncritical acceptance of a model of mind as a formal symbol system, a self-contained bundle of representations of the world and processes by which they are manipulated. This so-called representational view, in tandem with the computational theory of mind, was foundational to the development of artificial intelligence and by the mid-1970s essentially defined cognitive science (e.g., Newell & Simon, 1976). The "representations over computations" doctrine that continues to be embraced by the self-proclaimed cognitive science orthodoxy (e.g., Adams & Aizawa, 2008) has entrenched conceptual and methodological implications for cognitive psychology. Chief among them are a view of cognition as isolated from other human faculties and a focus on the representational states of the

---

[16] See, for example, Evnine (2008); Harré (1992); Tissaw (2010); and Tolman (1998).

individual mind (processing system). Investigatory strategies typically aim to decouple cognitive processes to discern their individual effects, with a view to articulating internal structures and mechanisms underlying cognitive states, which is accomplished most efficiently with controlled laboratory research design.

Yet that there is growing discontent with the orthodox view is highly evident, as illustrated in the stated aims of a newly launched series entitled *New Directions in Philosophy and Cognitive Science* edited by John Protevi:

> As cognitive science continues to make advances, especially in its neuroscience and robotics aspects, there is growing discontent with the representationalism of traditional philosophical interpretations of cognition. Cognitive scientists and philosophers have turned to a variety of sources – phenomenology and dynamic systems theory foremost among them to date – to rethink cognition as the direction of the action of an embodied and affectively attuned organism embedded in its social world, a stance that sees representation as only one tool of cognition, and a derived one at that (Protevi, 2007).

A similar review was offered by Anderson (2003):

> For over fifty years in philosophy, and for perhaps fifteen in Artificial Intelligence and related disciplines, there has been a re-thinking of the nature of cognition. Instead of emphasizing formal operations on abstract symbols, this new approach focuses attention on the fact that most real-world thinking occurs in very particular (and often very complex) environments, is employed for very practical ends, and exploits the possibility of interaction with and manipulation of external props. It thereby foregrounds the fact that cognition is a highly *embodied* or *situated activity* – emphasis intentionally on all three – and suggests that thinking beings ought therefore be considered first and foremost as acting beings (p. 91).

Though relatively new to artificial intelligence, the focus on *activity* and *practice* as analytic units has deep roots in earlier psychological and philosophical critiques of and alternatives to representationalist accounts of mind-world relations. By representationalist we refer to a doctrine holding that formal translations of the world into symbols are filed in an individual's brain/head and that these translations are the knower's point of access to that world. Among psychologists, Brentano's act theory (1874/1995), with acts understood as the means by which objects (world) are grasped (apprehended), and Dewey's (1930) attack on the "spectator view" of knowledge provide historical examples of alternatives to representationalism.

Yet in examining these examples and many that follow, we find several related and, frankly, fuzzy categories – acts, action, activity, practice. The terms are used freely, often loosely, and neither their demarcation from one another nor their philosophical origins as a set are settled matters. A rough classification of act and actions as solo performances and of activities and practices as collective is tempting, especially because the roots of "act" and action have been located in Aristotle (Anscombe, 1966) and of "activity" in Marx (Engeström & Miettinen, 1999). Indeed a distinction between act and activity is the basis of Leont'ev's three-level model of activity (1978), which Engeström and Miettinen (1999) called a "breakthrough." Related strategies for distinguishing action from activity focus on the goal-directed nature of the former versus the object-directed nature of the latter, again supporting a distinction between action as an individual level of analysis and activity as a collective-level notion (see Engeström, 1999).

Yet this classification is complicated by several counter-examples, including the "whole social act" that Mead discussed in the *Philosophy of the Present* (1932) and elsewhere, Dewey's analysis of technology, and the more recent incorporation of speech acts into activity theory, drawing especially from Wittgenstein's *Philosophical Investigations* (1953; see Engeström & Miettinen, 1999). Note that even Vygotsky argued that speech (the vehicle of participation in the social world) and practical action "are part of one and the same complex function, directed toward the solution of a problem at hand" (1978, p. 25). This functional relation is tied to the integral relation of (inter)action and situation, as expressed here by Dewey in *Experience and Education* (1938a, pp. 43–44):

> The conceptions of situation and interaction are inseparable from each other. An experience always is what it is because of a transaction taking place between an individual and what, at that time, constitutes his environment, whether . . . persons with whom he is talking . . . the subject talked about . . . the toys with which he is playing, the books he is reading or . . . an experiment he is performing. The environment, in other words, is whatever conditions interact with personal needs, desires, purposes, and capacities to create the experience which is had.

Thus the distinction between acts and activity is a blurry one at best. Clearly the concepts are intertwined and separable only artificially, and both overlap substantially with the more recent concept of practice (discussed later).

Action/Activity as Expressing Coordination and Overcoming Dualisms

Of greater import is that the psychologists Dewey, Vygotsky, and Leont'ev target acting and *activity* as principal units of analysis in an attempt to capture the coordinated operations of persons in the world that are the "psychical reality" the discipline seeks to understand. Even in Dewey's critique of the reflex arc concept, we see the act portrayed as a coordinated event, as he charged that psychology treats the reflex arc not as "a comprehensive, or organic unity, but a patchwork of disjointed parts, a mechanical conjunction of unallied processes" (p. 358). The "defect" in psychology's theoretical treatment of the reflex is the assumption that "sensory stimulus and motor response as distinct psychical existences, while in reality they are always inside a coördination and have their significance purely from the part played in maintaining or reconstituting the coördination" (Dewey, 1896, p. 360). Peter Godfrey-Smith has effectively summarized the integrated nature of the unit of analysis central to Dewey's concerns:

> For Dewey, when we want to make philosophical claims about thought, mind, or intelligence, we must take as the natural "unit" for analysis a large, inclusive structure.... The "unit" for discussion in philosophy of mind and epistemology is larger in two ways. First, mind is something that only exists in a social medium of symbol use. And more important for our present purposes, we must treat thought as something connected to distal environmental conditions at both ends. Dewey has an especially "wide" version of what is sometimes called "wide psychology." Thought is a response to specific kinds of environmental conditions, and the function or role of mind is to guide the agent in the physical transformation of those conditions, via action. As a consequence of this, all of Dewey's claims about the role of thought, or mind, or knowledge in the world are claims about a system that includes action and its consequences.... For Dewey, the role of thought and knowledge is not to passively reflect a pre-existing world, but to *change* reality. The outcome of a successful inquiry is a *transformation* of the situation (Godfrey-Smith, 2002, pp. S27–28).

The subtext to the emphasis on coordination is an assumption that through a turn to action/activity we are able to overcome dichotomized conceptions of subject and object, mind and world, body and mind:

> To be able to analyze such complex interactions and relationships...there is a demand for a new unit of analysis. Activity theory

has a strong candidate for such a unit of analysis in the concept of object-oriented, collective, and culturally mediated human activity, or activity system. Minimum elements of this system include the object, subject, mediating artifacts (signs and tools), rules community, and division of labor (Engeström & Miettinen, 1999, p. 9).

More recently, authors such as Papadopoulos (2008) have drawn on Vygotsky (1978) to argue that we cannot understand the volitional, cognitive, and emotional dimensions of practice as isolated but rather must examine their concrete and everyday realization in human action and community life.

The inherent coordination implied by the concepts of action and activity stands in stark contrast to the great bulk of psychological theorizing, in different eras and across research specialties. It contrasts with the elements of consciousness sought by introspectionists (e.g., Titchner), the stimulus–response mappings of behaviorists, the isolation of cognitive functions and mechanisms, and the identification of neural pathways evidenced with varying degrees of emphasis at different points in psychology's relatively short history as a formal discipline. Because in each case the intent of at least some authors was to frame the whole of psychology as under the covering laws of the dominant framework (e.g., structuralism, behaviorism, cognitivism), the temptation is to read these efforts as offering integrated general psychologies. However, in each case the unit of analysis is what has been hailed repeatedly by critics as a "reduction": A behavior is not an integrated unit of analysis, nor is a stimulus–response connection or a mechanism of cognitive or neural variety. The idea of "paradigms" of psychological thought confiscated clumsily from Kuhn is not equivalent to an integrated analytic starting point. Moreover, a frequent charge is the elimination of human agency and intentionality (albeit thorny problems) in any effort to streamline the analytic focus to a single category.[17] For example:

> The exceedingly broad use that psychologists have made of key terms and categories, such as behavior and learning, also has figured large in this overall reductionistic scenario. If all of human activity is understood as a kind of behavior independent of the intentions and circumstances of psychological individuals, then such behavior can be more easily equated with the non-intentional activity of say, for example, rats in mazes (Martin, Sugarman, & Thompson, 2003, p. 23).

[17] Held (2007) provides a comprehensive review of recent critical statements; see also Martin, Sugarman, and Thompson (2003); Robinson (1989); and Rychlak (1997).

Yet for the authors just quoted, reform is not to be achieved by the facile assertion of agency into psychological accounts. Intentionality itself must be understood as embedded in and enabled by inextricably social and material milieus: "The normative and perspectival character of human mentality reflects the reality of the human condition" (Martin et al., p. 99).

Acting, in requiring intentionality, implicates persons. One option, perhaps the only one, is to embrace, as did Stern, the (goal-directed) *acting person* as the fundamental unit of analysis. Tolman (1998) traced the idea to Aristotle and Hegel that "acting," even though attributed to a person, has an irreducibly social connotation as well, by virtue of which it is fundamentally connected to the moral order. In addition to the many philosophical expressions of a fundamentally social acting person, Tolman noted the less explicitly ethical versions of this same idea in the psychology of 20th-century figures including Vygotsky, Lewin, Mead, and J. J. Gibson. Drawing on these traditions, Martin, Sugarman, and Thompson noted that "to act is to make a choice, and choice cannot escape the question of value – that is, of what is the societally right or moral choice" (Martin et al., 2003, p. 13).

Note also philosopher Edward Pols' account of the inherently integrated nature of acting in the introduction to his treatise on agency, *Acts of Our Being*:

> When we act, some thing comes into being: in the first place, our act itself; in the second, ourselves, for in some measure we come into being by virtue of our acts. So at least it seems, as we experience ourselves as agents-in-act, as we complete and possess this experience in a rational web of words, and there is nothing so important to our ontological self-respect as the conviction that goes with it, not perhaps of creating being, but at least of participating as coadjutors in a coming-into-being. . . . our minds and bodies move in the order of the act (Pols, 1982, p. 1).

Some contemporary social theory, reflected, for example, in discursive psychology, introduces a distinction between acts and actions that is important to the notion of the acting person being forwarded here. Actions are "behaviors that are intended performances," and acts are the social meaning of actions (Louis, 2008, p. 25). Emphasis on the act*ing* person encompasses both the intentional quality of action and the social meaning or force of acts accomplished through the actions, for the intentional performances of persons (actions) always take place within socially negotiated or inherited contexts of social meaning. Furthermore, there is an imaginative dimension to meaning-making (sense-making) activity; imagining is a form of activity and imagination opens new ways of acting. Yet agency is expressed and meaning is made creatively always through the linguistic resources available in a given linguistic community (see Brockmeier, 2009). The acting person as an analytic unit then integrates intentionality, creativity, and social normativity. Against this understanding of activity, Percy Bridgman's remarks on science as activity are particularly cogent: "Science is activity. Science does not begin until my activities begin" (Bridgman, 1950, p. 50).

Several recent authors including William Smythe (1998) and Mark Bickhard (2008) have undertaken the attempt to offer the acting person as central to psychological analysis. As Smythe pointed out, "the concept of person has never received much sustained systematic treatment by psychologists, and is becoming increasingly problematic in contemporary psychology, with its steady advance toward depersonalized views of its subject matter" (1998, p. xi). He characterized the threat to a vigorous person-centered psychology coming from both cognitive neuroscience and extreme versions of cultural/constructionist psychology that view the person as a "faceless node in a network of social and cultural relations" (p. xii).

More central to the analysis of rationality and problem solving, Jean Lave (1988) positioned "persons-acting" as key to her analysis of cognition in the practice of "just plain folks" and as part of her argument for situating cognition.

Here we might pause to acknowledge that any claim, including our own, in which acting persons are the fundamental units of analysis in any endeavor begs questions about the nature of personhood, individuality, and agency. On these questions there is a vast philosophical literature beyond the scope of our discussion. Yet that these philosophical concerns are inescapable we readily acknowledge, and we do not consider the burden of bearing their weight sufficient reason to turn away from the pursuit of adequate integration in psychological accounts of science.

SCIENCE AS ACTIVITY AND PRACTICE

In the second half of the 20th century, the widespread turn to *practice* as a central unit of analysis in social theory and philosophy bears some important similarities to the earlier turn to *activity* in psychology and is indeed overlaid on the concept of practical activity (e.g., Garfinkel & Sacks, 1970). In his introduction to *The Practice Turn in Contemporary Theory*, Schatzki and colleagues began with the acknowledgment that "thinkers once spoke of 'structures,' 'systems,' 'meaning,' 'life world,' 'events' and 'actions' when naming the primary social thing. Today, many theorists would accord 'practices' a comparable honor" (Schatzki, Knorr Cetina, & von Savigny,

2001, p. 1). Engeström and Miettinen (1999) distinguished activity from practice principally by academic subdivision, with the situated learning contexts of education and the distributed cognition approaches in cognitive science roughly equivalent to the practice concept in science studies.

In recent decades, the term "practice" is more visible and influential than "activity," particularly in relation to science. Analysis of practice has been fruitfully appropriated in science studies to account for the inherent and inescapable complexities of science, including in the natural sciences.[18] Rouse noted,

> In the case of scientific inquiry, of course, the relevant practices include ways of encountering, responding to, and being resisted by what scientists are dealing with. The intertwining of language, social practice, and reality cannot be neatly bounded at the points where they run up against the natural world, for encounters with the world, and indeed the very boundaries between self and world, belong to interpretive human practices (Rouse, 2002, p. 85).

Although some analyses of science have limited conceptions of practice to what people do rather than anything involving "the head" (e.g., Latour, 1987), cognitive science increasingly has borrowed the frameworks and methods of anthropology to investigate "real-world" contexts in which cognitive practices such as reasoning and problem solving take place. Indeed, conceptual, discursive, or linguistic practices might constitute "the paradigm case of practices" according to Rouse, who has offered a practice-based philosophy of science. Moreover, the inclusion of "invisible activities" such as thinking was central to Dewey's critique of representationalism; indeed, thinking is a form of activity for Dewey (Dewey, 1926; Dewey & Bentley, 1949). William James also framed cognition, including theoretical, as act and activity:

> Cognition, in short, is incomplete until discharged in act; and although it is true that the later mental development, which attains its maximum through the hypertrophied cerebrum of man, gives birth to a vast amount of theoretic activity over and above that which is immediately ministerial to practice, yet the earlier claim is only postponed, not effaced, and the active nature asserts its rights to the end (James, 1896, p. 85).

In turning to practice as a theoretical unit, the emphasis is more obviously on the situated nature of cognitive activities. That conceptual and discursive practices are paradigmatic examples of practice underscores the important

---

[18] See Bourdieu (1990); Pickering (1987); and Rouse (1996, 2002).

element of normativity that must be central to elevate "practice" above the descriptive level of "what people tend to do" in everyday contexts. Although the term "practice" has been criticized as vague and insufficiently explanatory (Turner, 1994), Rouse has emphasized its philosophical status as normatively constrained performance (Heidegger, 1962; Wittgenstein, 1953), distinguishing it in this way from the looser notions of social regularity, commonality, or custom: "Not all practitioners perform the same actions or have the same presuppositions, but practitioners and other constituents of a practice are accountable for performances or presuppositions that are inappropriate or otherwise incorrect. . . . Such a network of practices need not be identifiable as regularities of action or belief, even as a whole" (Rouse, 2002, p. 168).

In science practice normativity is irreducibly both natural and social, as Rouse made clear in commenting on Kuhn's significance:

> Kuhn's concern was not whether there "really" are objects, natural kinds, or causal powers independent of scientific practices. Rather, his point was that what scientific practices are normatively accountable to are objects and phenomena manifest through scientists' engaged, meaningfully situated practices of experimentation and theorizing. Muon-detecting is in this respect no less "intersubjectively meaningful" than voting or bargaining, and grasping the significance of an experimental result need be no less affective than recognizing one's own humiliation (Rouse, 2002, p. 84).

In our analysis, we mean by "normativity" at a most general level that there are constraints on activity (and practice). Some constraints are imposed socially, in accordance with standards of procedure (method), productivity, and professional conduct; some are material, tied to the composition of objects, their properties, and relations; and some constraints are cognitive, imposed by the processes and structures by which we are able to categorize and make inferences. As we shall make evident in the course of our analysis, we also understand normativity to include forms of *demand*, including demands from the material environment of the scientist (e.g., to "care" for cells in order to keep them alive; see especially Chapter 4); social demands including justifying and marketing one's discoveries; and cognitive demands for order and coherence in experience. Because it foregrounds normativity, the concept of "practice," like that of activity, most adequately captures the coordinated functioning of individual and (social) domains, thereby circumventing the individual framework evinced by representational accounts. Indeed, Rouse distinguished practice from "mere activity" by appealing to the integration of the situation in which activity takes place: "Practices are not just agents' activities but also the configuration of the world within which those activities are significant" (1996, p. 133). Therefore, taking into account the inherent coordination of activity and action, and the normativity – irreducibly social, and material, and *cognitive* – that structures and constrains actions within traditions of practice, we offer the following unit of analysis for our psychological study of science:

> *The acting person in normatively structured contexts of practice within which sense-making and identity are two central tasks. These tasks are continuously negotiated in relation to other persons and with the objects and artifacts important to that context of practice.*

This is our conception of both science and psychology.

Qualifiers

The turn to acting persons in contexts of practice that we embrace as central to adequate accounts of both science and psychology is suggestive of a thoroughgoing pragmatism. Here we offer some qualification. Although there are many important respects in which the pragmatism of Dewey and Mead have informed our framework and the frameworks from which we draw (distributed cognition, situated cognition, embodied cognition), there are limits to this influence and important points of departure. Philosopher Marjorie Grene, whose stance as a foremost philosopher of biology is described as "personalistic and contextualist" (Auxier, 2002), offered a scathing but trenchant summary of pragmatism's potential shortcomings, noting that when her colleagues "mention James or Dewey, I simply scream or gnash my teeth":

> Pragmatism was directed against a number of things, but it contained nothing positive beyond the pleasant desire to make things comfortable. Whatever is uncomfortable-death, sin, despair – it passes by on the other side. As has been said a number of times, pragmatism is afraid to face evil. And it is afraid, too, to face the ultimate puzzle of human individuality. To be sure, the individual and the activities of the individual that pragmatism, like existential philosophy, is supposed to devote itself to. But it is the "adjusted" individual, the stereotyped individual who has forgotten how to be an individual that pragmatism celebrates (Grene, 2002, p. 11).

The emphasis on adjustment is the fallout of an emphasis on problem solving and functional adaptation in pragmatism, and the ghost of this stereotyped "adjusted" person lingers in the new frameworks drawing from Dewey's functionalist commitments, even if the person's acts are now adequately theorized as situated and embodied. In consequence, in applying this framework to science practice we risk too tidy a picture of scientific life. Indeed, this is a risk or risky consequence of the turn to practice in science studies in general. Importantly, moreover, when applied to human beings, theories centered on functional adaptation (as is implied by "acts" and "practices") do not sufficiently take into account the historical systemic forces that work to impede the adaptation of some subsets of persons in some cultures (e.g., women and minorities).

Yet though conceptually rooted in pragmatism (given the original meaning of *pragma* as action), contemporary activity theory and practice analysis in science studies do not assume a seamless, progressive process, and whether this is the best-fitting interpretation of Dewey is indeed debatable. Yet certainly as actions and activity or actions and practices are construed as interconnected, the story becomes more convoluted and less buoyant:

> Actions are not fully predictable, rational, and machine-like. The most well-planned and streamlined actions involve failures, disruptions, and unexpected innovations. These are difficult to explain if one stays at the level of actions. The analysis of the activity system may illuminate the underlying contradictions that give rise to those failures and innovations as if "behind the backs" of the conscious actors (Engeström, 1999, p. 32).

It is also helpful to remember the point made by William James to the effect that pragmatism is only a method, and not a very new one, having been used to varying degrees by Socrates, Aristotle, Locke, Berkeley, and Hume (James, 1907/2003).

SUMMARY AND ORGANIZATION OF THE BOOK

The central question prompting this text is how to characterize the actions of human beings at what most would consider the more rational end of their functioning: the practice of science. Paying homage to the traditional self-definition of human beings as essentially rational agents, we assume that the psychological understanding of science facilitates an understanding of humans more generally. Addressing two related concerns that we identify as the integration problem in science studies and the problem of grab-bag general psychology, we have been studying research scientists in biomedical engineering labs as they engage in formulating and solving cutting-edge problems in their respective domains.

Our unit of analysis for this project is *the acting person as scientist,* by which we intend to emphasize the coordination rather than isolation of processes: cognitive, social, cultural, emotional, agentive, and material. Nevertheless, as Anscombe's (1966) philosophical analysis of the concept of action clarified, people act under different descriptions of the action in question. Anscombe was referring to descriptions people give of their own actions, yet we benefit from the insight that, for the sake of analysis, it is frequently helpful to describe actions with different points of emphasis. We take for granted that the different aspects cooperate and, indeed, are mobilized in action itself: That is the point of considering this text an effort to address problems of integration and general psychology and the point of suggesting the acting person as the unit of analysis in science.

We concentrate in this text on two broad categories of action in science: sense-making and identity negotiating. Although sense-making has emerged as a well-defined theoretical construct in organizational research (Weick, 1995), we use the term as shorthand for the continual efforts of any person to sort, understand, plan, and evaluate experiences of every kind, thereby giving them meaning. Two aspects of Weick's understanding of sense-making are particularly germane to our project. First, our study highlights the irreducibly social *and* cognitive aspects of sense-making, which is consistent with our understanding of the laboratory as a cognitive-cultural system. Second, Weick has emphasized that no sense-making occurs without a sense-maker, thereby tying the activity of sense-making to an embodied and intentional actor and explicitly linking sense-making with identity formation. Therefore our categories of sense-making and identity negotiation can be distinguished only with great imprecision. Indeed, identity negotiation can be considered a form of sense-making directed to the meanings one applies to oneself within social groups that include but are not limited to the particular research laboratory, one's field of practice (biomedical engineering), and science as a tradition of inquiry.[19] Under sense-making we examine by turns the *problem-solving* person (Chapter 3) and the *feeling* person (Chapter 4) in laboratory practice. As we discuss in Chapter 4, feeling dimensions of laboratory practice are interwoven with those more traditionally called cognitive. Under identity we examine the *positioning* person (Chapter 5) and the person negotiating culture by means of

---

[19] An excellent review of identity theory is provided by Burke and Stets (2009).

enacting race and gender relations in the laboratory (Chapter 6). It is in relation to identity in particular that the principally adaptive nature of activity is called into question, as discussed in these chapters. In the context of discussing positioning we also consider attributions of agency to objects and artifacts that contribute meaningfully to the problem-solving act at hand. We then examine how all of these aspects of sense-making and identity formation come together in relation to a single acting person by examining the *learning* person (Chapter 7).

Across chapters there is overlap and cross-reference as is consistent with our theme of integration. The chapters analyzing categories of action are preceded by a discussion of our methods (Chapter 2), including an overview of the laboratory settings and the researchers therein, and a description of our efforts to make sense of the laboratories as a multidisciplinary team. We include both a brief review of other work that has been published from our study of these laboratories and a description of the principal methods of analysis used in the present project.

Chapter 8, the final chapter that we have titled "Epilogue," provides a discussion of the implications of our analysis and relates them to earlier efforts to construe science from the standpoint of the acting person, including those of D. L. Watson, George Kelly, Michael Mahoney, and Ian Mitroff. The point of the final chapter is that that there is a tradition of imagining science as psychology, if not one officially recognized. This book represents both an effort to make that tradition more visible and to offer a new contribution toward it.

2

Methods of Study

In this chapter we describe the contexts of science practice that we studied and the approaches to the data collection and analysis that are the basis of this text. Along the way we note methodological issues and controversies as they relate to our analysis and apply more generally to the use of qualitative methods in psychology. Thus this chapter is comparable to the methods section of a traditional research study, but it contains departures in form in keeping with the theoretical aims of this text.

PARTICIPANTS, CONTEXT, AND HISTORY

The Laboratories

Overview

Our framing assumption is that the cognitive practices of the laboratory are both *situated* in the laboratory and *distributed* across systems of interacting persons, artifacts, instruments, and traditions. The situated approach to cognition construes intelligent behavior as arising within particular settings such that its features are dependent on that setting, in contrast with a view of cognition as an abstract realm or self-regulating process. The assertion is that "problem solving is carried out *in conjunction with the environment*" (Brown, Collins, & Duguid, 1989, p. 36, emphasis added). By distributed, we mean that we regard brain and environment as co-constituting a single complex system, inasmuch as the forms of sense-making and problem solving that occur would not be possible in isolation from that environment, including the social environment. This situated and distributed framework employed in the analysis to date connects our study of the laboratories to other investigations of real-world problem solving that implicate the environment in cognition in an important way.[1] Although the meaning of "distributed cognition" as casting the unit of analysis of cognitive practice as comprising people and artifacts is clear, the word "distributed" is perhaps not the best choice. As Hall, Wickert, and Wright (2010, p. 2) have noted, "the word 'distributed' . . . is an adjective, past tense, and a modifier of already existing cognition." That is, it is not an indicator of process, but of already accomplished activity. If we want to understand and analyze cognition in practice – as it unfolds – we need to understand the acts of "distributing" cognition. That is, "distributing is a verb, operating in an ongoing present, and shifts our attention to studies of how cognition . . . is produced historically out of human activity." In previous work, our research group has emphasized that a laboratory might be understood not simply as both the physical space plus its collection of artifacts, such as instruments and specially designed technologies, but also as an organized social group with a shared agenda that undergirds the particular problem-solving goals undertaken by any researcher at any given time.[2] The principal investigator of each lab is most obviously involved in setting this agenda; however, our analysis has shown that the agenda is dynamically influenced by contributions from all members of the laboratory community. We thus construe the laboratory as a "problem-space" – comprising researchers and artifacts – with permeable boundaries, in that it enables researchers to move between its physical boundaries and the wider community to which the work is connected (Nersessian, 2006).

Our investigation is situated in two biomedical engineering (BME) research laboratories located on the campus of a major research university in an urban setting. Biomedical engineering may be characterized as an *interdiscipline*,[3] meaning that "melding of knowledge and practices from more than one discipline occurs continually, and significantly new ways of thinking and working are emerging" (Nersessian, 2006, p. 127). The two BME laboratories we target engage one form of *interdisciplinarity*, merging resources drawn from both biology and engineering in the form of researchers, concepts, and methods. In addition to the blending of academic domains, the labs tend to attract persons with diverse interdisciplinary interests and experiences. Indeed, the range of experiences brought into the lab

---

[1] For example, Greeno (1998); Hutchins (1995a, 1995b); Resnick (1996); and Zhang (1997).

[2] See Kurz-Milcke, Nersessian, and Newstetter (2004) and Nersessian, Kurz-Milcke, Newstetter, and Davies (2003).

[3] See Kurz-Milcke et al. (2004) and Nersessian (2006).

is broader than is indicated by the engineering backgrounds identified. For example, researcher A35 describes his background as combining biology, physics, chemistry, math, and computer programming.

The interdisciplinary nature of the work of these laboratories affects their social and learning cultures by effectively distributing knowledge and competencies. In turn, this distribution helps encourage researchers at all levels to seek help and apprenticeship as appropriate to the task at hand. All researchers, including undergraduate students, are encouraged to make important research contributions.

Notably, each laboratory is also characterized by its *innovation*-seeking agenda. Researchers in both labs actively seek new ideas and applications at the cutting edge or frontier of knowledge in their respective fields. These are therefore *creative* environments, which in previous work, Nersessian (2006) has characterized as providing a basis for distinguishing the study of these laboratories from other environments that are driven by problem solving in which the goal is not novelty but precision, such as Hutchins' studies of navigation processes undertaken in landing a plane or steering a ship to harbor (Hutchins, 1995a, 1995b).

The laboratories are also *evolving* systems, undergoing continual transformation.[4] That is, the social and epistemic culture (Knorr Cetina, 1999) of the laboratory transforms in response to the activities of its researchers, including the entry of new researchers and the departure of others, and to collaborations formed with other laboratories through these transitions. These changes alter not only the social composition of the laboratory but also its knowledge resources and the finer points of its agenda – the projects and problems under concern. Technologies are developed or modified in accordance with new problems and applications. Thus the historical embedding of each laboratory and its ever-changing web of conceptual and practice relations to the field of which it is a part contribute to the understanding of the laboratories as evolving problem-spaces and evolving learning cultures.[5]

We investigated university research laboratories in tissue engineering and neural engineering. We collected data in each lab over a period of 2 years, with 2 years of follow-up collection as needed. In the next two sections we provide further details relevant to each laboratory and its specific research agenda.

---

[4] See Nersessian (2005, 2006) and Nersessian, Kurz-Milcke, Newstetter, and Davies (2003).
[5] See Kurz-Milcke et al. (2004) and Nersessian, Kurz-Milcke, Newstetter, and Davies, (2003).

Lab A

Lab A is a tissue engineering laboratory that dates to 1987 and was established by the principal investigator who continues to direct it. During our study, the main members included a director, one laboratory manager, one post-doctoral researcher, seven PhD graduate students (three graduated while we were there, and the other four graduated after we concluded formal data collection), two MS graduate students, and four long-term undergraduates (two semesters or more). Of the graduate students, two were male and seven were female; the postdoctoral researcher was female. Additional undergraduates from around the country participated in summer internships, and international graduate students and postdocs visited for short periods. The laboratory director (A13) was a senior, highly renowned pioneer in the field of biomedical engineering. All of the researchers came from engineering backgrounds, mainly mechanical or chemical engineering, and some were currently students in a BME program. Some had spent time in industry before joining the lab. The lab manager had an MS in biochemistry. Researchers frequently consulted with a histologist located in the building and traveled to other institutions for various purposes, including to collect animal cells and to run gene microarray analyses.

Lab A's overarching research problems are to understand the mechanical dimensions of cell biology, such as in the behavior of endothelial cells in response to shear forces, and to engineer living substitute blood vessels for implantation in the human cardiovascular system. The dual objectives of this lab explicate further the notion of an engineering scientist as having both traditional engineering and basic scientific research goals. Examples of intermediate problems that contributed to the daily work during our investigation included designing and building living tissue – "constructs" – that mimics properties of natural vessels; creating endothelial cells (highly immune sensitive) from adult stem cells and progenitor cells; designing and building environments for mechanically conditioning constructs; and designing means for testing the construct's mechanical strength.

Usually the graduate student researchers work on individual projects, often with assistance from undergraduates. As one example, an undergraduate researcher (A35) noted that his main task was assisting a PhD student in her project, which was trying to improve the mechanical strength of constructs. As part of his effort to assist her, A35 added *decrin*, a protein that has an affinity for collagen. He used histology to see how the collagen and decrin aligned and used a mechanical tester to examine the effects of decrin on construct strength. As another example, A5 provides this summary of her research: "*I examine endothelial cell genetic behavior under fluid shear stress conditions correlating to the development of atherosclerosis using tissue engineered vascular grafts as a physiologic model. I also profile behavior of endothelial cells and their progenitors for use in lining tissue engineered vascular grafts*" (from the researcher profile).

Lab D

Lab D is a neural engineering laboratory. During our study the main members included a director, one laboratory manager, one postdoctoral researcher, four PhD graduate students in residence (one left after two years, and three graduated after we concluded formal data collection), one PhD student at another institution who periodically visited and was available via video link, one MS student, six long-term undergraduates, and one volunteer for nearly two years, who was not pursuing a degree (already had a BS) but who helped out with breeding mice. Of the graduate students, two were female and three were male. The postdoc was male. When we began, the laboratory director (D6) was a new tenure-track assistant professor, who had done a postdoc in a biophysics laboratory where he worked on developing techniques and technologies for studying cultures of neurons. He already had achieved some recognition as a pioneer. His background was in chemistry and biochemistry, with his engineering knowledge largely self-taught, though highly sophisticated. The backgrounds of the researchers in Lab D were more diverse than those in Lab A and included mechanical engineering, electrical engineering, physics, life sciences, chemistry, and microbiology; some were currently students in a BME program. As an institution, the neural engineering laboratory had been in existence for only a few months and was still very much in the process of formation when we began data collection.

Lab D's overarching research problems are to understand the mechanisms through which neurons learn in the brain and, potentially, to use this knowledge to develop aids for neurological deficits and, as the director of Lab D, D6, put it, "to make people smarter" (PI). Examples of intermediate problems that contributed to the daily work included developing ways to culture, stimulate, control, record, and image neuron arrays; designing and constructing feedback environments (robotic and simulated) in which the "dish" of cultured neurons could learn; and using electrophysiology and optical imaging to study "plasticity." Here, again, the researchers have dual scientific and engineering agendas. Because all the projects centered around the "dish," there was significantly more interaction among research projects than we witnessed in Lab A.

Unlike the traditional independent configuration of Lab A, Lab D is embedded in an open space that is shared by seven faculty members and their postdoctoral researchers, as well as graduate and undergraduate students.

DATA COLLECTION AND ANALYSIS

Overview of Issues Posed

The present effort to analyze science as psychology is grounded in a wider laboratory study directed by Nersessian and Newstetter and an interdisciplinary team of researchers,[6] but veers from it somewhat in focus and methodology.

We begin with some brief background comments about the use of qualitative methods in psychology before turning to a description of those methods we particularly have engaged. Many volumes are written on these topics, so it is not our task to devote much space to the controversies here. However, they lurk in the background of any psychological analysis and therefore must be identified and owned rather than left to fester in the shadow of our positive claims.

Problems related to the conceptual integration of psychology noted in Chapter 1 overlap with concerns about integration in methods, yet the methods provide a set of conceptual quandaries of their own. Indeed, the methods used to address psychological questions have received far more attention historically than have the questions themselves or the concepts undergirding the questions.

In psychology, particularly in North America, the acceptance of qualitative data and analysis as on equal footing with quantitative has been slow,

---

[6] Research on the labs discussed in this book began in 2001 and ended in 2009. It was conducted under the leadership of Nancy Nersessian and Wendy Newstetter. In seeking to describe and understand the learning, reasoning, and problem-solving practices of both the novice and expert researchers in interdisciplinary laboratories, Nersessian and Newstetter were guided by the goal of trying to account adequately for the contributions of both culture and cognition to laboratory practices and products. The framework builds on Nersessian's previous work interpreting the role of conceptual models in the reasoning practices of scientists and in science learning (e.g., Nersessian, 1992, 1995, 2002). The works produced to date through the laboratory study reflect the diversity of interests and distribution of expertise. A wide range of publications by the Cognition and Learning in Interdisciplinary Cultures research group can be found at www.clic.gatech.edu.

despite recognition that the distinction between qualitative and quantitative ultimately is artificial. Qualitative approaches in all their great variety remain in the background of psychological methods, as a defiant minority. However, it is likewise evident that, across the social sciences and even within psychology itself, the visibility and influence of qualitative approaches have been increasing. More dramatically, Smith described the "explosion of interest in qualitative psychology" as a significant shift in a discipline that has hitherto emphasized the importance of quantitative methodology" (Smith, 2003, p. 1). The various effects of this shift include the inclusion of qualitative PhD theses, the publication of qualitative research in peer-reviewed journals, and courses in qualitative methods added to the requirements or at least electives for psychology majors.

Yet qualitative analysis and interpretive methodologies remain steeped in controversy and ambiguity. Challenges to the legitimacy of these approaches as a foundation for knowledge and questions concerning the generalizability (and hence, usefulness) of the analytic products they yield continue to keep qualitative psychologists on the defensive with their more traditional disciplinary colleagues. Among the reasons for the hesitant embrace of qualitative strategies by psychologists is that much of what passes for procedure entails seemingly irreducible acts of insight; thus much is not amenable to description, let alone replication. Qualitative analyses have difficulty passing reliability tests established for the purpose of evaluating quantitative data, prompting charges that qualitative analysis represents "mere storytelling" rather than a solid descriptive foundation.

However, these charges reveal fundamental misconceptions about science process, within which qualitative analysis plays an essential role in every phase of procedure. Contemporary efforts devoted to the *study* of science, including cognitive studies of science, inherently make use of case- or problem-based analysis of practices in context, for which control is decidedly not the goal and quantitative comparison constitutes merely one item in a well-stocked toolkit.

Yet it is perhaps incumbent on qualitative researchers to demonstrate the cross-cutting relevance of their approaches. One strategy for doing so is to identify more appropriate and cogent metaphors for the projects of qualitative analysis. For example, Alasuutari's metaphor of *unriddling* has application to both science and the humanities:

> Any single hint or clue could apply to several things, but the more hints there are to the riddle, the smaller the number of possible solutions. Yet each hint or piece of information is of its own kind and equally

important; in unriddling – or qualitative analysis – one does not count odds. Every hint is supposed to fit in with the picture offered as the solution. (Alasuutari, 1995, p. 7)

However, this very example underscores a profound problem that complicates the process of demonstrating the relevance of qualitative research in psychology. The unriddling metaphor and the notion that analysis is a solution, even if one of many possible solutions, suggest an underlying realist epistemology that is not shared by many qualitative researchers. Many see the project of qualitative analysis as one of giving voice to participants, particularly voices historically stifled through methodologies more traditional to the discipline of psychology (e.g., Gergen, 2000).

Questions concerning the object of analysis correspond to the controversy concerning the proper construal of the role of the qualitative researcher. Some authors have explicitly identified a tension between the idea of researcher as instrument and the goal of producing rigorous and trustworthy interpretations (Poggenpoel & Myburgh, 2003). That is, whether the particularity of the researcher, not just that of the participant, is to be central or minimal to the analysis is a matter of the theoretical commitments undergirding the particular project undertaken: "Depending on the underlying paradigm, we may work to limit, control, or manage subjectivity – or we may embrace it and use it as data" (Morrow, 2005, p. 254).

Why is this problematic and not merely cause to celebrate the potential for dialogue and exchange? Writing in 1971 on the logic of naturalistic inquiry, Norman Denzin acknowledged that "the basic unit of naturalistic inquiry has never been clarified and the role of the naturalistic observer in his studies remains clouded" (p. 166). Little has changed in the ensuing decades, despite more widespread acceptance of qualitative methods among social scientists. Currently, competing approaches to qualitative analysis can serve as a bewildering deterrent to psychologists. In part the variety of named approaches represents a response to criticism and suspicion. That is, the qualitative research community has sought increased legitimacy for interpretive approaches through achieving greater specificity in relation to the name of the procedure followed for a particular act of interpretation. Conventions are in place to keep researchers on board with the particular contours of sanctioned analytic trends, the rules and boundaries of which are less easily specified than those of traditional quantitative methods in psychology. Thus it is not enough to claim that one is using qualitative, even ethnographic analysis. One must be able to identify a particular form or school to which analytic efforts conform. For example, if we say that we have used grounded theory, we call into question whether it is the version favored by Glaser (1992) or Strauss (1987; Strauss & Corbin, 1998). Or if we are using discursive analysis, we must name the variety to which we aspire.

These comments are in no way intended to minimize the importance of distinctions in qualitative methodologies used in psychology or in the social sciences more broadly. For of course epistemological commitments intertwine with method to an inseparable degree. By aligning with Strauss or Glaser we position the logical foundation of our work as principally inductive or not. By subscribing to Harré's discursive analysis (Harré, 1998; Harré & Gillett, 1994), we channel the later Wittgenstein; with the Leicester School, we channel Foucault (Edwards, 1997). Phenomenological analysis cannot be severed from the broader effort to establish Husserl and Heiddeger as foundational to human science (Giorgi, 1970). Thus more important than the name for the purposes of accountability is the unit of inquiry and of analysis, as Denzin (1971) acknowledged. Here one must make an ideological commitment to a basic question, problem, or analytic focus, which then dictates the particular method used, rather than base the decision on the mechanics of the process that most obviously fits the procedural steps taken.

In the introductory chapter of this text we identified our theoretical unit of analysis as *the acting person*, the scientist, *in normatively structured contexts of practice*, namely the science laboratory. This translates method-ologically into an analysis of *acts of coordination* achieved in the context of the biomedical engineering laboratories. Although we include ethnographic and cognitive historical analysis in developing conceptual categories, for our purposes the acts most exemplary of coordination are discursive strategies, specifically as recorded in the context of individual interviews with labora-tory researchers and in group meetings they held for their own purposes.

The question might well be raised why we focus on *interview text* rather than video recordings of laboratory practices. In the learning sciences, the use of video recordings is seen as central to the empirical analysis of complex interactions of persons with one another and with the objects of their practices; video recordings enable consideration of the interrelations of verbal utterances (talk), gestures, use of tools and artifacts, and both routine and novel practices (Jordan & Henderson, 1995). Although we have collected numerous videotapes for the very purpose of analyzing the interaction of researchers and their artifacts, in this text we are specifically concerned with the problem of understanding the sense-making and identity *achievements of participants within the interaction*, in addition to the interaction itself. The focus on interaction itself, as Jordan and Henderson (1995) readily admitted, comes with an assumption that knowledge is a fundamentally social product. We worry about the possibilities of eliminating the affective, motivational, and cognitive particularity of contributors to the collective practice of knowledge construction through accounts that risk a kind of social reductionism. We have no easy solution to the problem of adequately understanding the contribution of the particular to the collective without resorting to an individualistic framework, but the inclusion of the personal dimension of science is necessary if we are to ever move past the received artificial separation of the social and cognitive realms that has dominated accounts of science to date and that contributes to the ongoing problems plaguing psychology. There can be no final means of solving this problem, and of course we do not offer a resolution here.

Our working strategy is to combine a framework of distributed and situated cognition with a methodology of analyzing individual interviews, which we characterize as discursive events or productions. That is, we approach our study of these laboratories with the assumption that cognitive and social processes constitute a whole. Yet through the use of individual interviews with researchers with different levels of expertise and from different disciplinary backgrounds, we are able to analyze how the particular learning history, relational networks, and affective style of the researcher might contribute to the rich mélange of social *and* cognitive practices that constitute science. We do not assume that the individual interview provides us with a telescope into the inner world, the private experience of participants. However, we recognize that the narrative provided by an interview (in this case a "situated interview" in that it takes place in the environment – the laboratory – in which the cognitive activities of interest occur) provides the best data available for analyzing these aspects that have tended to be left out of integrative efforts to date. The discursive strategies we chose to present here are examples; thus the task is to descriptively illustrate the coordination rather than to demonstrate or explain it causally.

The discursive events on which we rely most heavily are interviews with researchers that we collected over the course of several years, from 2002 to 2006. Several members of our group became participant observers of the day-to-day practices in each lab, and we estimate that in total we spent more than 800 hours in the labs. We have used our observational findings together with the interviews to arrive at our interpretations of lab practices. We took field notes on our observations, audiotaped interviews, and video- and audiotaped research meetings (full transcriptions have been completed for 148 interviews and 40 research meetings). Indeed, we have collected far more data than we will ever be able to analyze given the constraints on our time and resources. Therefore the work is in progress, and what we offer here should be read as a pause in the process of analysis rather than as a definitive statement.

One problem that arises in relation to the use of interview data is what has been termed by ethnographers as the "say-do" problem, meaning that what people say they do in describing their practices is not necessarily an accurate reflection of those practices. The corresponding assumption held by some researchers is that ethnographers are better able than the participants themselves to describe practices through the rigorous application of observation and collaborative production of hypotheses that can be tested against further data (Jordan & Henderson, 1996). Although we cannot overcome the say-do problem entirely, two points are relevant here. First, many conceptual categories arising from the interviews, including those relating most directly to problem solving and emotion, were corroborated through laboratory observation of practices. However, some categories, such as those relating to identity, cannot be observed directly but require interpretation from researchers' talk about their work. Second, our analysis focuses on the *function* of various statements (strategies) within discursive events (interviews) as much as on the content. More specifically, we are concerned with the psychological function of statements within the interview context. In some cases we assume these to have implications for the psychological function of certain words or patterns in the laboratory practices more generally (e.g., the widespread practice of using anthropomorphism in relation to cells in Lab A and cell networks in Lab D.) Because we are not directly asking about the psychological categories that are the focus of our analysis here, the problem of matching speech to practice (or saying to doing) is less cogent than it might be when describing practices for the sake of better understanding the particular science itself.

One strategy we did not include that might have been a valuable addition is to analyze the function of discursive strategies in recorded conversations between researchers in each lab, or in the course of meetings between researchers, as Kevin Dunbar has employed in his studies of analogy use in laboratory settings (Dunbar, 2001). However, we consider the researcher's account of his or her own practice to be an invaluable aid to understanding the psychological dimensions of science practice. The interview provides insights into how each scientist understands her work, what it means to her, and how she experiences it. Moreover, following Joseph Rouse (1996), we regard the interview as a conversation that is itself part of the process of science, rather than as a representation of the actual processes of problem solving as they are lived in the laboratory. That is, the need to painstakingly explain their work to interviewers from outside their field has been described by some of our participants as contributing to new ways of framing and making sense of what they are doing for themselves. The PIs of both labs noted that our interactions have made their researchers more reflective.

Therefore, we use ourselves and our collaboration as analytic instruments, relying on the basic human capacities of insight and argumentation as we engage with the accounts of our participants. This is in keeping with our claims about the nature of the science we study. It would be absurd to consider the acting person any less the means than the object of our analysis. Ortner's summary of the process is instructive:

> Ethnography, of course, means many things. Minimally, however, it has always meant the attempt to understand another life world using the self – as much of it as possible – as the instrument of knowing. As is by now widely known, ethnography has come under a great deal of internal critique within anthropology over the last decade or so, but this minimal definition has not for the most part been challenged. Classically, this kind of understanding has been closely linked with "fieldwork," in which the whole self physically and in every other way enters the space of the world the researcher seeks to understand. Yet implicit in much of the recent discussions of ethnography is something I wish to make explicit here: that the ethnographic stance (as we may call it) is as much an intellectual (and moral) positionality, a constructive and interpretive mode, as it is a bodily process in space and time (Ortner, 1995, p. 173).

Thus ethnography and qualitative analysis, more generally, entail inherently integrated acts of collection and analysis. The very questions guiding the methodology selected reflect the social and academic background, disposition, and style of each researcher and the particularities generated by the groups that researchers form for collaboration. Constraints and affordances of opportunity (funding, time, personnel, and other resources) are interacting factors. Only within and through this rich flux can acts of interpreting and justifying interpretations be understood adequately.

Consistent with this view, our approach to understanding the practices of biomedical engineers embraces a range of methodologies reflective of the academic discipline, experiences, and "style" of the authors, each of whom represents a unique configuration of interdisciplinary interests and influences. We provide here a description of our processes of data collection and analysis, both for the purposes of accountability and in the interests of facilitating further study of this kind.

DATA COLLECTION PROCEDURES

Ethnographic Observation

Before conducting interviews, ethnographers on our research team (including the principal investigators, postdoctoral research scientists, visiting faculty, and both graduate and undergraduate cognitive science students) spent many hours in the laboratories observing and informally interacting with laboratory researchers. Each ethnographer then "hung out" in a lab, scheduled observation or interview sessions, and attended laboratory functions (meetings, presentations, dissertation defenses) in accordance with the number of hours he or she was officially devoting to our project and the specific goals associated with the observation.

Field notes were collected primarily by the two principal investigators and student members of the research team, each of whom kept regular hours in the laboratory for the purposes of observing and engaging in informal, spontaneous conversation with laboratory participants. In addition, we collected video- and audiotapes of journal club and other team meetings held by researchers in each of the labs.

Members of the research team involved in collecting data with researchers in both laboratories conducted the *interviews*. Each interviewer independently scheduled and made arrangements for the interviews he or she conducted. Thus for some participants interviews occurred at regularly scheduled intervals, usually every 2 weeks, and in other cases the schedule was more sporadic. Most interviews took place within the laboratories, invoking the space with reference to the objects and representations (e.g., microelectrode array [MEA] recordings of network activity as projected on a computer screen) relevant to the problem or issue under discussion. In total, we conducted 72 interviews with researchers in Lab A and 75 with research staff in Lab D.

The interviews are of different types and have different foci. All interviews are unstructured, but most have specific aims. Some are aimed at obtaining the lab members' accounts of their learning and social experiences since entering the laboratory, including their involvement in mentoring relationships; some are focused on their personal history before they entered the laboratory; some have the objective of understanding the development and current use of technology and equipment specific to the laboratory; and most are focused on the research problems the researchers were then working on and how they relate to the larger laboratory goals.

We were also given guided tours of the laboratory, which constituted something between an interview and a field observation. We audiotaped and transcribed these tours.

Gender and Race Enactment Data Collection

Malone's interviews focused more explicitly on race and gender enactments, and participants were drawn from Labs A and D as well as several additional engineering laboratories on the same campus. Given that "identity"-related codes were emerging from the grounded theory analysis described earlier, we considered it important to work with a rich conception of identity, which would include the social categories of race and gender. The method used to address race and gender questions was not grounded theory per se, but rather an analysis that drew more explicitly from the psychoanalytic perspective of Jacques Lacan and from social theory relating to lived race and gender experiences. Within Labs A and D Malone conducted research on gender over a 10-month period; it involved 10 intensive interviews with lab members as well as observations and recordings from the lab and from lab meetings. Although questioning was open ended, the questions asked related to particular themes: how one handles failure, how one "becomes" part of the lab, how it feels to be in a lab with so many women, and what participants like about biomedical engineering.

For the study on race, conducted over an additional year, 24 participants, all from underrepresented minority groups and all female, were drawn from biomedical engineering and other sciences. Malone conducted individual interviews with 15 participants, of whom 5 agreed to participate in a second interview. She thus ultimately had 20 individual interviews. During the same period, five focus groups were also held. They were publicized by email and each ran for 1 hour, with lunch provided. Given our initial focus on gender, the focus groups were limited to women. Students dropped in and out of the focus groups although there was a core group who attended most of the meetings.

In the 20 individual interviews, the interviewer used an ethnographic approach aimed at understanding the social reality of the interviewee and the cultural situation of her practices (in this case race/minority status within a White-majority school science lab). Although conducted with an initial set of prepared questions with a variety of aims, the interview protocol was individually tailored, and there was a dynamic relationship between the answer and follow-up questions. This procedure accords with ethnographic interview protocol as outlined by Spradley (1979). First, the research was explained, pointing out current data on the status of minorities in science, and the interviewer (Malone) spoke of our group's specific interest in research labs. Second, in light of the sensitive nature of discussing race in a situation where careers were at stake and the interviewer was White, she always began with a "grand tour" question where the student was asked to describe her history and interest in science, her lab, and then her experiences in the lab. These descriptive questions helped establish rapport and also served to help ground our later questions. Follow-up questions were developed in advance in relationship to the themes that emerged in the focus groups and a review of the current literature.

Transcription and Notation

For both laboratories, we assigned aliases, sanitized transcripts, and kept taped conversations secure to protect the confidentiality of participants. All interviews were transcribed by either the interviewer or student members of the research team hoping to gain experience in ethnography. Interviews and field notes were archived in an extensive database designed and managed by a student member of our research team to whom we are particularly indebted.

We used the following convention to label and catalog interviews: year-month-date-Lab (A or D)-i (for interview)-participant # (e.g., [2006–10–20-A-i-A22]). Of note in relation to the presentation of interview segments appearing in this text, a marked change in direction or pauses in speech are indicated by a dash (–). The authors' condensation of interview text is indicated by (...). For Lab tours, the notation differs slightly: year-month-date-Lab (A or D)-Tour (e.g., [12–02–2002-D-D6 (Tour)]).

Triangulation of Data

Our interdisciplinary investigatory team held regular weekly meetings, during which we compared ethnographic observations as recorded in field notes with the content and emerging interpretations of the interviews. We hammered out and refined emerging conceptual categories during these meetings. As noted, the evolving composition of the team affected both the style of working together and the specific categories that emerged or received emphasis. Codes that emerged through grounded theory analysis (described later) were "tested" for their applicability and conceptual fit with data recorded as field notes and with a sample of additional interviews.

CODING AND ANALYSIS

Grounded Coding

Our process was broadly informed by grounded theory (Strauss & Corbin, 1998) in the sense that we endeavored to use constant comparison and analytic induction to develop coding schemes and conceptual categories to characterize the laboratory practices we observed and the discursive productions obtained through interviews. Within this framework, however, much variety is allowed and enacted in accordance with the preferences and disciplinary emphases of the interpreter, as well as the specific questions posed. Moreover, the line between grounded theory and theory-led thematic analysis becomes blurred as analysis continues and hypotheses are formed from initial coding. Therefore it is most accurate to say that we used a variety of interpretive strategies, including analysis of the function of language within the discursive event of the interview (discourse analysis), interpretive content analysis, case study analysis, and cognitive-historical analysis. As the investigation proceeded, we became increasingly interested in the lived relations among researchers and between researchers and the artifacts central to their practice (see Nersessian, 2006; Osbeck & Nersessian, 2006).

The remaining chapters in this book highlight not only different content foci (i.e., different categories within a general psychology) but also reflect our range of interpretive strategies and guiding questions, with each chapter illustrating a subtly different approach. In addition to reflecting our full embrace of the researcher-as-instrument guideline, we believe that the combination of methods facilitates the integration of cognitive, cultural, and affective accounts of science practice.

All chapters except those addressing race and gender draw from coding done across transcripts in a manner consistent with grounded theory and interpretive content analysis. With an overarching goal of inclusiveness, two researchers (Osbeck and Kurz-Milke) selected a subset of interviews that included representatives from both labs with varying levels of research experience, academic levels, and disciplines. These selected interviews were analyzed progressively, line by line, from beginning to end, with the aim of providing an initial description for most if not all of their passages. This step is akin to the open coding phase of grounded theory, wherein the goal is to take the data apart and begin to characterize it in terms of conceptual or descriptive dimensions. From these initial descriptions, codes were developed in an effort to thematically organize descriptions, and these tentative codings were discussed in research group meetings.

Two main approaches to developing themes/codes became apparent: 1.) sampling across interview texts (both within each lab and across labs) and 2.) focusing on the experience of one participant over time (case study). Both approaches were consistent with grounded theory approaches that emphasize the emergence of themes (and codes) from described units of text. However, the first approach was the initial strategy used, and some of the codes developed through this strategy informed the case study analysis. The higher order categories, that is, emerged from the analysis within and across labs. For example, the theme of seeking coherence, particularly in relation to the notion of distributed demands for such coherence, emerged from descriptions of passages that suggested a kind of latching onto or alignment with something established. Similarly, the theme of narrative seemed to serve the same function of achieving coherence through different means (in this case, telling a story).

The highest level codes that emerged were *Seeking Coherence (sense-making)*, which included subcategories of modeling, framing, positioning, and offering narrative (lab history and personal history); *Identity; Affect/Motivation; Agency; Norms; Desires,* and *Pragmatic Focus.* The idea for this book grew out of the fact that many of the emerging categories seemed to depart in theme from the original goals of understanding the problem solving practices of interdisciplinary researchers, most notedly, the tools, artifacts, and representations used for knowledge construction. We did not anticipate coding for affect or identity, for example. In this sense we were surprised by some of the codes we found useful in describing the content of some passages. The range of codes reminded us of a general psychology text; hence, the title and direction of this book.

Throughout the coding process, codes were generated by formulating a succinct and thoroughly plausible description of a selected passage, given its particular context within the interview and to some extent the larger context of the laboratory's workings and recent history. These descriptions were arrived at through detailed discussion about the possible significance and alternative interpretations of the text. At this stage in the analysis, which was tied as closely as possible to interview data, developing original, accurate, and specific descriptions took precedence over concerns about the number of codes and descriptions used. In other words, we made no attempt to minimize the number of coding categories at this stage.

The recurrence of codes in the interview analysis had the effect of ensuring their further discussion and reevaluation. A code was recorded only when both researchers were in full agreement about its fit and relevance to the given passage. We revisited descriptions and codes throughout the process in keeping with new thinking and discussion of text, as well as new observations from the laboratory. This step in our procedure is very like axial coding in grounded theory, wherein the goal is to refine categories and begin to make connections among them. At this stage, many codes were refined, involving further distinctions. Some were revised, some added, and some eliminated as thinking evolved. As one example, a distinction between "model-based understanding" and "model-based reasoning" emerged on the basis of the analysis of verb use ("reasoning" being assigned to passages of text in which references to "it seems" or "it might be" were foregrounded in conjunction with an organized representation). We analyzed the assigned codes for conceptual similarity, overlap, and distinction.

We then grouped codes together under headings that seemed to capture as much as possible their important main theme. For example, model-based understanding and model-based reasoning were included under the superordinate heading of "Model-based cognizing," along with model based-description or explanation, which seemed to express situations in which the model was invoked principally for the purposes of expressing a concept to the interviewer. Codes that did not fit easily into one of the main headings were analyzed further for possible overlooked meanings or their fit with other categories. This process was repeated until no further reductions could comfortably be made. We then developed and revised a written description of main code categories, with examples of text passages assigned to each category. Main categories, descriptions, and examples were brought to the main research team for feedback and were revisited and in some cases revised after the feedback was received.

As noted, in addition to sampling across texts, another strategy used was to focus coding and analysis on interviews with one particular lab member over time. Some of the interviews were conducted at regular 2-week intervals: Chapter 7, "The Learning Person," illustrates a kind of case study analysis that analyzed chronologically one researcher's developmental trajectory from a point very soon after she first entered the laboratory. Here we also used a coding system similar to grounded theory in that the codes emerged from the careful line-by-line reading and interpretation of the interview text, and some of the content was assigned to categories guided by questions that specifically concerned learning.

Rigor and Plausibility

Although fully embracing the idea of researcher as instrument and thus trusting in the insights gleaned from careful reading of the text, we sought rigor in coding and analysis in three principal ways, which can be thought of as three phases of accountability:

1. *Collaborative coding.* Although as noted, each of the authors concentrated on a different aspect of analysis, coding and other forms of analysis (e.g., interpretations and categorization of field notes) were initially conducted by at least two members of the interdisciplinary team working in collaboration. This ensured that any codes assigned seemed reasonable and important to at least two persons familiar with the laboratories and the aims of the overall project. Some of these coding sessions were recorded so the process could be evaluated during external audit.

2. *Group feedback.* Updates on coding were presented at regular research meetings of the interdisciplinary research team. At times some rather heated arguments between researchers with different disciplinary backgrounds resulted. Codes were retained, adjusted, or abandoned in line with feedback from these sometimes difficult group meetings.

3. *External audit.* We hired an outside evaluator: an assessment specialist working within the university in which the laboratory research was conducted. The auditor was familiar with qualitative methods, but was not involved with the project itself. His task was to assess the coding process for rigor of procedure and to closely examine the codes for plausibility and fit. We received a very favorable report on both the process followed and the codes produced through our collaborative efforts.

ADDITIONAL METHODS OF ANALYSIS

Cognitive-Historical Analysis

To date, ethnography has been the primary method for investigating situated cognitive practices. As a method, it does not, however, suffice to capture the critical *historical* dimension of the research lab: the evolution of technology, researchers, and problem situations over time that is central in interpreting the practices. To capture the "evolving" dimension of the lab we used a mixed-method approach that adds to ethnographic methods those of cognitive-historical analysis.

In this context, cognitive-historical analysis uses historical and philosophical methods of analysis to examine historical records with the objective of understanding cognition, rather than constructing a historical account per se. It is a case study method that follows trajectories of the human and technological components of a cognitive system on multiple levels, including their physical shaping and reshaping in response to problems, their changing contributions to the models that are developed in the lab and the wider community, and the nature of the concepts that are at play in the research activity at any particular time.[7] As with other cognitive-historical analyses, we used the customary range of historical records to recover how the representational, methodological, and reasoning practices have been developed and used by the BME researchers. The practices can be examined over time spans of varying length, ranging from shorter spans defined by the activity itself to spans of decades or more. For this dimension of the part of our study, we collected the following types of data: publications, grant proposals, dissertation proposals, PowerPoint presentations, laboratory notebooks, emails, materials related to technological artifacts, and interviews on lab history.

Cognitive-historical analysis interprets and explains aspects of these practices in light of salient cognitive science investigations and results (Nersessian, 1992, 1995a, 2008a). Saliency is determined by the nature of the practices under scrutiny. Although some researchers including Nersessian (e.g., Gooding, 1985; Tweney, 1989) have used the method to construct richer, more nuanced interpretations of historical episodes, in the context of our research on the labs, the objective of cognitive-historical analysis is not to construct a historical narrative. Rather, the objective is to enrich understanding of cognition in context through examining how knowledge-producing practices originate, develop, and are used in science and engineering domains.

Psychoanalytic Analysis

The analysis of race and gender enactments in relation to scientists' identity formations drew more explicitly from a framework informed by the psychoanalytic theory of Jacques Lacan, which is the expertise of Kareen Malone.

---

[7] For a comparison of cognitive-historical analysis to other methodologies in cognitive studies of science, see Klahr and Simon (1999).

Unlike the rest of our categories, race and gender were explicitly addressed in the context of the interviews, and the interviews might be said to pivot around their enactment in laboratory practices. By contrast, as noted, interviews from which the other analytic categories emerged were organized around problem solving or equipment use; conversation was not explicitly aimed at extracting themes relating to emotion or positioning, for example. The emotionally charged nature of the categories of race and gender and the degree of sensitivity around the questions relating to these categories in science practice called for a different approach to data collection and analysis than was appropriate for the bulk of our investigation. Speech remained the focal point; thus interviews and focus group transcripts were used as the principal tools. However, one would expect that in addition to what was said, there was a great deal worth paying attention to in what was *not* said.

Because the analysis of gender undertaken in this work is informed by psychoanalysis, the analysis of gender reflects a way of reading and a way of listening. The listening is something that necessarily happens in the actual interview; it is constrained by the relationship implied by this sort of research. The reading, always done with another researcher, entails an awareness of places of contradiction, slips, odd grammatical phrasing, wherever there is an *elsewhere* in the text or what Lacan (1998) called an "inter-dit" (a between meaning). Close reading occurs at points where words seem to stray from their intended meaning, where there might be a rupture in the usual concordance with others and with "reality," and where contradictions seem important. Interpretations that are derived from a psychoanalytic reading may be interrelated with other thematic findings; for example, around questions of failure.

SUMMARY

Our working alliances have been made more invigorating if not always easier by differences in background and style. Despite the differences in emphasis and the ways they are reflected in our methodologies, we have also benefited from a good deal of cross-pollination through our ongoing discussions and negotiations. We are familiar with and respectful of one another's ways of working with the material and have attempted to find common ground and learn from one another as we wrestle with our large dataset and the enormous complexity of the integration we seek to describe.

3

The Problem-Solving Person

We begin our investigation of the research laboratories by focusing on the overarching purpose for which they were created: to solve complex, interdisciplinary problems. Activities surrounding problem solving drive much of what transpires in these research labs. It has long been a central assumption of cognitive studies of science and technology that the cognitive resources scientists bring to bear in problem solving are not different in kind from those used in more ordinary instances, but rather lie along a continuum. Construing the problem-solving strategies that scientists have developed as sophisticated, highly reflective outgrowths of ordinary reasoning and representational practices allows researchers in this area to both draw from and inform the study of the nature of cognition.

The idea that problem solving plays a significant role in cognitive processes has been central to cognitive psychology since its emergence in the mid-20th century. Earlier, however, the notion that problem solving is an important part of the processes involved in learning, creativity, insight, and cognitive/conceptual change was represented in Dewey's analysis of the essential elements of effective pedagogy in *How We Think*:

> The best, indeed the only preparation is arousal to a perception of something that needs explanation, something unexpected, puzzling, peculiar. When the feeling of a genuine perplexity lays hold of any mind (no matter how the feeling arises), that mind is alert and inquiring, because stimulated from within. The shock, the bite, of a question will force the mind to go wherever it is capable of going, better than will the most ingenious pedagogical devices unaccompanied by this mental ardor. *It is the sense of a problem* [emphasis added] that forces the mind to a survey and recall of the past to discover what the question means and how it may be dealt with (Dewey, 1910, p. 207).

Of particular note in this passage from Dewey is the unmistakable relation of affect to the experience of problem solving, a topic we engage more fully in Chapter 4. More generally, knowing is for Dewey a process or activity that necessarily implicates world and actor, rather than something that will result in an acquisition (knowledge). As an activity, knowing consists of transforming a problematic situation into a resolved situation, which frequently involves hands-on manipulation of the problem-space. This transformation occurs primarily through engagement in deliberation, calculation, reflective manipulation, and experiment.

European psychologists of the Würzburg and Gestalt schools of psychology also demonstrated concern with problem solving as a central feature of intellectual functioning, as did Jean Piaget. Herbert Simon, a founder of cognitive science, is well known for his early conception of cognition as problem solving (Newell, Shaw, & Simon, 1958). He traced this facet of his intellectual lineage to a member of the Würtzburg school, Otto Selz. Simon famously proposed that the kinds of problem solving implicated in scientific discovery "are not qualitatively distinct from the processes that have been observed in simpler problem-solving situations" (Simon, Langley, & Bradshaw, 1981, p. 2). However, the notion of scientific change as a problem-solving process cut across several schools of psychological thought, as well as the history and philosophy of science as practiced in Europe early in the last century – most significantly by the historians Emil Meyerson and Alexandre Koyré and the philosopher Karl Popper, who studied in the Würtzburg school. These strains of psychological, historical, and philosophical thought had a considerable influence on each other, as discussed in the brief account by William Berkson and John Wettersten (1984), which focused on the relation of the Würtzburg school to Popper's philosophy of science.

In keeping with the traditional cognitive psychology framing, we conceive of problem solving as a form of information processing that uses representations and reasoning in pursuit of goals. However, our research aims at an integrated study of the embodied, artifact-using, and socio-cultural dimensions of human cognition – one in which the environment does not simply *scaffold* cognition, but is integral *in* cognitive processes. Thus, consistent with contemporary "environmental perspectives" on cognition (Nersessian, 2005), we frame the problem-solving person as embedded in complex cognitive-socio-cultural systems and examine problem-solving processes as situated in specific contexts and as comprising persons and select artifacts. Note that the environmental perspective we embrace makes real-world analyses of problem solving – that is, problem solving in natural settings of practice – much more relevant to our interests than controlled experimental studies of scientific reasoning. Much of the psychological literature on problem solving focuses on "well-structured problems" for which the representations, procedures, and goals are clearly defined, such as solving a standard algebra problem. By contrast, the kind of problem solving our biomedical engineering researchers engage in is "ill-structured" in that these aspects are vague, open-ended, and continually evolving, with many potential pathways to solution.

The frontier nature of the problems formulated within the laboratories has led us to characterize these labs as *innovation communities*. Their broad objectives are (1) to create new knowledge at the intersection of biology and engineering, such as understanding the biological properties of endothelial cells in terms of the mechanical forces they experience as blood flows through the artery, and (2) to create new engineering artifacts composed of – or interfacing with – biological materials, such as a blood vessel implant engineered from living tissue. Designing, building, and experimenting with physical simulation models are central problem-solving practices in the engineering sciences. Model-based simulation is an epistemic *activity* that involves experimenting; generating and testing hypotheses; providing explanations; and drawing inferences.

What is striking is that all aspects of the systems in each innovation community (laboratory) – human and technological – are concurrently undergoing change, reconfiguring the system as the research program moves along. Although there are loci of stability, artifacts and understandings undergo transformation over time. Learning is ubiquitous. As we have found, the technologies and the researchers have evolving, intersecting trajectories that need to be factored into an account of the problem solving at any point in time. Coupling ethnographic and qualitative methods of analysis with methods of cognitive-historical analysis has enabled us to examine in a unified manner the evolution of problems, technologies, and learning and how these are enacted in daily problem solving.

AN INTEGRATIVE SCHEME FOR ANALYZING PROBLEM SOLVING

The problem-solving practices of engineering scientists provide an excellent resource for advancing the agenda of environmental perspectives because their simulation technologies perform simultaneously as cognitive and cultural artifacts. As a way of capturing relations among cognitive and cultural facets of problem solving, we formulated an analytical scheme comprising cognitive, investigative, and interactive practices: Cognitive practices give rise to and interact with investigative practices, which in turn are enacted through interactional practices, which support and sustain both. Separating these aspects for analytical purposes is not meant to imply that they are separated in practice; to the contrary, they are deeply interwoven in research processes. In this scheme, cognitive practices include the formulation of problems and objectives, reasoning practices, and approaches to problem solving that drive the work in the lab. Deriving from these practices, the investigative dimension comprises the nature of the lab members' daily work as they strive to develop understanding, techniques, and technologies. The interactional dimension comprises the lab culture and the social configurations – the kinds and nature of relationships formed in the lab and how they are sustained. In this chapter we examine these interwoven facets by drawing on our field observations and the expressions in the interview transcripts we have coded as relating to the main problem-solving practice in both labs: simulative model-based reasoning, by which we mean reasoning by constructing and manipulating models (Nersessian, 2008a).

The central cognitive practice in both of the laboratories is *traversing the in vivo–in vitro divide*. Research in biomedical engineering often confronts the problem that it is both unfeasible and unethical to carry out experiments directly on animals or human subjects. Thus, ways must be devised to emulate selected aspects of in vivo phenomena to a degree of accuracy sufficient to warrant (to the extent possible) transfer of simulation outcomes to the in vivo phenomena. As a result, the central investigative practice common to both labs is to design, build, and experiment with in vitro simulation models that *parallel* selected features of in vivo systems. As one respondent noted,

> We typically use models to predict what's going to happen in a system [in vivo]. Like people use mathematical models to predict... what's going to happen in a mechanical system, well this is an experimental model that predicts what would happen – or you hope it would predict – what would happen in real life [2006–05–23-A-i-A7].

The researchers refer to their physical models as "devices," which are usually signature artifacts of that laboratory or of a small group of laboratories in the research area. Experimentation with devices requires configurations that bring biological and engineering components together in an investigation, which the researchers refer to as a "model-system":

> When everything comes together I would call it a 'model-system.'... I think you would be very safe to use that [notion] as the integrated nature, the

*biological aspect coming together with an engineering aspect, so it's a mul-
tifaceted modeling system. I think that's a very good terminology to describe
that* [2002–07–09-i-A-A10].

Devices and model-systems are structural, behavioral, or functional
analogue representations of in vivo phenomena of interest. They are also
engineered systems, with design and material constraints that can impose
simplifications and approximations unrelated to the biological systems
they model. As we will see, to engage in new problem solving, researchers
need to understand in what ways these physical simulation models are
simplifications or approximations and, to some extent, the history of the
design of the devices within the lab.

Model-systems, as *hybrid* objects, reflect the laboratories as bioengineering environments. They combine biological materials, such as cells and
collagen, with engineering components, each possessing their own affordances and limitations. Given the hybrid nature of the simulation models,
a major learning challenge for researchers is to develop selective, integrated
understandings of biological concepts, methods, and materials and of engineering concepts, methods, and materials. By "selective," we mean that a
researcher-learner needs to integrate, in thinking and experimenting, those
dimensions of biology and engineering relevant to their research goals and
problems. For example, in Lab A, researchers need to develop an integrated
understanding of the endothelial cell in terms of the stresses of fluid dynamics of blood flow in an artery. In Lab D the notion of plasticity, a biological
property, needs to be understood in terms of quantitative measures based
on recorded spikes of electrical activity of the neuronal network.

Many interactive practices develop around the devices. Building activities
that are centered on devices are pervasive in both laboratories, and they
serve several functions within the research community. Devices connect the
cognitive practice of in vitro simulation with social practices; for instance,
much initial mentoring and learning of the laboratory ethos take place in the
context of cell culturing. For newcomers, building cell cultures, constructs,
and microelectrode array (MEA) dishes provides opportunities to quickly
become absorbed into the culture of the lab community. As learners mature
into full-fledged researchers, they come to see the simulation models they
design and build as partners and active participants in their research. We
discuss this interactional practice, which we call "cognitive partnering,"
later in the chapter. First, we examine some of the model-systems as they
figure in the problem-solving practices of the laboratories, followed by a
discussion of our interpretations of the problem-solving person as enacting
these practices within distributed cognitive-cultural systems. We provide

extended descriptions of central devices and model systems here so we can simply refer to them in later chapters.

PROBLEM SOLVING IN A TISSUE ENGINEERING LABORATORY

The assumption that guides research in Lab A is that mechanical forces in/on the cardiovascular system have biological effects (most notably on the endothelial cells that line blood vessels) and contribute to disease processes. The lab director noted that starting in the late 1960s he began to realize that

> *characteristics of blood flow* [mechanical forces] *actually were influencing the biology of the wall of a blood vessel. And even more than that. . . . the way a blood vessel is designed is – it has an inner lining called the endothelium. It's a monolayer. . . . it's the cell layer in direct contact with flowing blood. So it made sense to me that, if there was this influence of flow on the underlying biology of the vessel wall, that somehow that cell type had to be involved, the endothelium* [2001–09–19-i-A-A13].

Therefore, understanding the biology and addressing problems of disease processes require an account of these dynamics. Over the years the research has advanced to now having as its ultimate applied objective the development of artificial blood vessels, engineered out of living tissue with the characteristics appropriate to functioning within the human body. Such an artificial vessel needs to be of sufficient strength to withstand the powerful forces of blood flow and to be lined with endothelial cells that are able to proliferate. The daily research in Lab A is directed toward formulating and solving problems that are smaller pieces of the grand objective. In the next sections we discuss two of the central devices that form the major components of model-systems in this lab.

The Flow Loop

Lab A has been in existence since 1986 and was created specifically to move the research "*from animal studies to cell culture.*" The lab director described the motivation for this move:

> *We felt the environment was too uncontrolled in a living animal. The endpoints we were looking at were ones that took place over 24 hours. There was no way to have constant flow for 24 hours. Blood flow changes not only beat to beat, but also when you are eating there's a redistribution of blood, more going to the gut. When you're sleeping there's a redistribution of the blood* [2001–09–19-i-A-A13].

So the first problem was to design and construct a means of simulating blood flow over the cells such that controlled studies with shorter durations could be carried out. Additionally the earlier studies of flow patterns and cell morphology had to be conducted separately and consolidated after the fact. The line of research with *cell cultures* – endothelial cells and smooth muscle cells – provided the basis for the initial configuration of Lab A. In the cell-culture line of research, an engineered artifact – a flow channel with accompanying flow-inducing components (called the "flow loop" in Lab A) – serves as an in vitro model paralleling certain in vivo conditions of the blood vessel, including both normal conditions and the pathology that previously was induced in living organisms. The cells are fixed to slides, placed in the engineered flow channel, and exposed to shear stresses from a fluid with the viscous properties of blood. The flow loop pump reflects knowledge of how fast blood flows in vivo, and the pulse dampener turns the flowing liquid from a pulsating flow to a smooth flow that allows control over the constancy of flow. These in vitro experiments on the response of the cells to shear stresses are based on an established fluid dynamic model; specifically, the fluid mechanics of a long channel with a rectangular cross-section. Using this device, changes in cell morphology (elongation and orientation) can be related directly to the controlled wall shear stresses.

To be able to make inferences about the outcomes of simulations and experiments with the devices, the researchers need to be aware of the various constraints that inform the design and construction of the models. This awareness is evidenced in numerous interviews. For instance, A10 expressed the researchers' general concern that

> as engineers, we try to emulate that environment, but we also try to eliminate as many extraneous variables as possible. So we can focus on the effect of one or perhaps two, so that our conclusions can be drawn from the change of only one variable [2002–07–09-i-A-A10].

Therefore, this is the central question: What is essential for a device to function as a useful model and what is "extraneous?" As a model, the flow loop represents the shear stresses during blood flow in an artery to a "*first-order approximation of a blood vessel environment . . . as blood flows over the lumen*" and enables "*a way to impose a very well-defined shear stress across a very large population of cells such that their aggregate response will be due to [it] and we can base our conclusions on the general response of the entire population*" [2002–07–09-i-A-A10].

Yet even though a flow loop simulation is dynamic, as a model, it does not represent the diachronic nature of the in vivo environment. For instance, blood flow in vivo changes when eating and sleeping. Therefore, eliminating the confounding factors leads to a simulation of the in vivo environment that is, as another researcher noted,

> *something very abstract because there are many in vivo environments and there are many in vivo conditions within that environment. Things change constantly in our bodies over our lifetimes; including physiological flow rates* [2003–04–29-i-A-A23].

From the outset, redesign has been a central activity within Lab A. The flow loop provides a major instance of this activity. In working with cultured cells, contamination is a constant problem, and it was the driving factor in the redesign of the flow channel device. An interview with a former graduate student, who at the time of the interview was a successful faculty member at another institution, elaborates on this problem and the subsequent redesign:

> *So, when I got here in 1994, the flow chamber was a mess. It was a bench top system, it had bulky tubes that looked something like some time machine from the 1950s or something. . . . But anyway it was quite messy and you know culture studies have to be done at 37 degrees so the way that they would do this was you know, incubators were certainly around in 1994, they would wrap these coils, these heating coils around these glass reservoirs and because it had to be a set flow, they would use a hydrostatic pressure difference to derive the flow, and a clamp, a regulated clamp to try and regulate the flow through the chamber and out into the – into the lower reservoir. So you had two reservoirs, one at the top, and one at the bottom, there'd be a hydrostatic difference between them, and then things would flow and then this whole thing would be sitting on the bench top – big bulky glass reservoirs with bulky tubing. . . . And this was subject to about a 50% success rate.*

Interviewer: In terms of contamination?

> *In terms of contamination. And the reason was because this whole thing had to be assembled outside of the hood [colloquial for "the sterile workbench"]. There was no way you could assemble this thing to stand up – this thing was on stands – you have to assemble this part outside of the hood, so basically we would connect these joints here, and connect them outside of the hood. . . . Doing experiments longer than 48 hours was almost impossible, because at experiments longer than 48 hours the incidence of contamination was probably greater than 90%. . . . I really like compact designs. . . . I instituted a lot of the things I saw over there [referring to the lab at which he had interned] in our laboratory, and one of the things was model-revising this design to go into the incubator. And, that was really why we moved*

> *from a system that required heating coils and an upper reservoir and a*
> *lower reservoir to a system that was just flow driven with a peristaltic pump*
> *and a pulse dampener that was – and everything could be done inside the*
> *incubator with smaller tubing, little reservoirs as opposed to big reservoirs*
> [2003–4–29-i-A-A23].

"*Model-revising this design,*" as the former graduate student described his contribution to this line of research, means redesigning the physical system, its parts (e.g., the reservoirs, the tubing), set-up (e.g., on stands in the lab vs. compact and in the incubator), and the physical principles governing its functional design (e.g., hydrostatic pressure difference vs. integration of a peristaltic pump). The actual flow channel, which is the part where the liquid flows over the cell cultures, was left untouched in this particular redesign, though minor changes have since been made to accommodate the introduction of a new device, the vascular construct model (described in detail later). Even though, in this case, redesign did not involve those parts where the cells-in-culture interfaced with the mechanical device, the revision of the flow loop was central to its function as part of a model-system, the success of which is dependent on solving the problem of contamination of the cell cultures.

Over time, new simulation devices have been designed, in particular, a "wet" device that is locally called "the construct." It provides a living cell wall model and has the potential not only to aid in understanding mechanical forces creating pathology but also in providing a tissue-engineered vascular graft that might someday repair diseased arteries.

The Construct

The cells that form the endothelium, a monolayer of cells that make up the inner lining of the blood vessel, are a major target of study in Lab A. In vivo these cells are in closest proximity to the blood flow. To the extent that cells are required to survive and perform in particular ways, the culturing of cells needs to emulate the naturally occurring conditions of living tissue in an organism. This emulation requires such factors as having the appropriate CO_2 and temperature levels in incubators. Until the late 1990s flow loop experiments were done only on cell cultures on slides, which were usually coated with a substance to make the cells adhere. After the flowing process, the cells were removed and examined by means of various instruments such as the Coulter counter and the confocal microscope to determine the effects of the mechanical properties of the flow on cell shape (morphology), alignment, proliferation (reproduction), or migration (locomotion). How-ever, as the lab director observed, *"cell culture is not a physiological model,"* although *"it is a model where biologic responses can be observed under care-fully designed and well-defined laboratory conditions."* The original flow loop was designed to "condition" cells, and though much still is learned by just using cells, in the late 1990s the director decided to set a major new direction for the lab:

> To use this concept of tissue engineering to develop better models to study cells in culture. Putting cells in plastic and exposing them to flow is not a very good simulation of what is actually happening in the body. Endothelial cells, which have been my focus for 30 years, have a natural neighbor called smooth muscle cells. If you look within the vessel wall you have the smooth muscle cells and then the inside lining is the endothelial cells, but these cell types communicate with one another. So we had an idea: Let's try to tissue engineer a better model-system for using cell cultures [2001–01–19-i-A-A13].

And so, shortly before we entered the lab, the researchers took what he called *"the big gamble"*: creating a living model of the blood vessel wall. To address this problem, they used biological and engineering materials and methods to design and create what they call "the construct" device. Although originally conceived as a potentially *"better model-system for using cell cultures,"* it came to be thought of as having the potential to develop into a blood vessel implant for human cardiovascular system repair.

As noted, the construct model marks a move toward a more physiological model – one whose function along mechanical, physical, and biochemical dimensions is more like the in vivo vessel. Building this model is a highly complex undertaking. An actual blood vessel has a tubular shape and com-prises several layers: the lumen where the blood flows; a first monolayer of endothelial cells that sit on collagen; an internal elastic lamina; a second layer of smooth muscle cells, collagen, and elastin; an external elastic lamina; and a third layer of loosely connected fibroblasts. In vivo, the cells create an extracellular matrix that is a network of proteins and other molecules that provides growth factors and mechanical properties. The kinds of problem solving that Lab A members engage in to refine this model involve highly sophisticated bioengineering, such as creating their own collagen gel tech-nology or determining whether it is possible to create a reliable source of endothelial cells from stem cells or adult progenitor cells.

Unlike cell cultures on slides, constructs are three-dimensional surfaces in which cells are embedded. In the in vitro culturing process, the construct is seeded onto a tubular-shaped silicon sleeve. In building a construct different kinds and degrees of approximation are used depending on the nature of the experiment. For instance, it is possible to use only collagen and not add elastin. Most often the cells are porcine or bovine, not human aortic endothelial and smooth muscle cells. Some experiments are conducted with only a single layer of the blood vessel wall with either endothelial cells or smooth muscle cells. Thus "the construct" forms a family of models, each designed in the context of a specific experimental purpose.

As with the flow loop, in our interviews the researchers provided evidence of understanding the constraints and affordances of the construct as a model. Most significantly, when used as a model-system with the flow loop, the constructs need to be cut open to lie flat within the flow chamber due to the design of the flow loop. This is understood to be an appropriate approximation of the tubular surface of an in vivo blood vessel. However, because the cells are so small with respect to the blood vessel wall, the flatness is not an approximation for the main objects of study – the cells. That is, from the "*cell's perspective*" it is not an approximation because, as noted by A10,

> The cell sees basically a flat surface. You know, the curvature is maybe one
> over a centimeter, whereas the cell is like a micrometer – like 10 micrometers
> in diameter. It's like ten thousandth the size, so to the cell it has no idea that
> there's actually a curve to it [2002–07–09-i-A-A10].

That is, flowing liquid over these flat constructs is thought to provide sufficiently accurate data on the responses of the endothelial cells that line the arterial wall, because the cells are so small with respect to the arterial wall that their "perception" of the wall is as flat as in the in vivo situation.

Another issue is that the medium flowing over the construct lacks a vertical component (flow is two-dimensional) and is unidirectional, whereas in vivo "*it sloshes around in the blood vessel.*" Yet again the focus is on first-order effects unless there is evidence of a need to consider higher order effects, such as if they find "*that there's a whole different pattern of genes that are up-regulated in pulsatile shear, or something, maybe then it would be more interesting to use different constructs and stuff.*" Under such conditions, a new flow-loop–construct model-system would need to be designed to simulate higher order effects.

Although the introduction of the construct has led to new devices and thus new model-systems, various configurations of flow-loop–cell or flow-loop–construct model-systems predominate problem solving in the lab.

A Vascular Construct Model-System Configuration

In experimental situations models tend to be put into configurations; that is, they are not isolated objects, but stand in particular relations to other models – researcher conceptual models and artifact models. Analyzing the design and execution of an experiment provides a means of articulating this feature of models. Soon after A7 arrived at Lab A she was designated the *"person who would take the construct in vivo,"* meaning that her research was directed toward conducting experiments with an animal that serves as a model for the human body in the context of the experiment. This objective immediately required that she would (1) need to design and build a construct that would both more closely mimic the functional characteristics of an in vivo artery than was used in most other experiments and would have sufficient strength to withstand the force of in vivo blood flow, (2) modify the flow loop so that it would work with constructs in tubular form, and (3) arrange for an animal (baboon) to be surgically altered so as to be able to experiment with the construct outside of its body and in a minimally invasive way. It also required that she bring together the strands of research being conducted by nearly all the other lab members. As she expressed it, *"To go to an in vivo model we have to have all, well most of the aspects that people have studied."* She proceeded to identify components stemming from the work of other researchers, stressing *"so this is integration of what they've learned. . . . You can't just study one aspect. . . . To improve mechanical integrity . . . is gonna obviously integrate the results of colleagues here in the lab"* [2002–06–17-i-A-A10].

Analyzing the development of her relationships to other researchers within the laboratory would require a chapter in itself. Instead, we include here a diagram of the research space of Lab A as drawn at our request by the lab director; it clearly positions A7 as central to most of the research in the lab (see Fig. 3.1).

When we entered the lab, A7 had been there about a year, but was still in the process of defining the specific goals and problems of her research. Our analysis is based on field observations of her conducting the parts of her research that took place in the laboratory (the baboon lived in a laboratory in another state); interviewing her as her research progressed, culminating with a final postgraduation interview; video and audio transcripts of the lab meetings in which she presented her work; and several of her writings.

A7's final overarching formulation of the problem was to determine whether it would be possible to use circulating endothelial cells

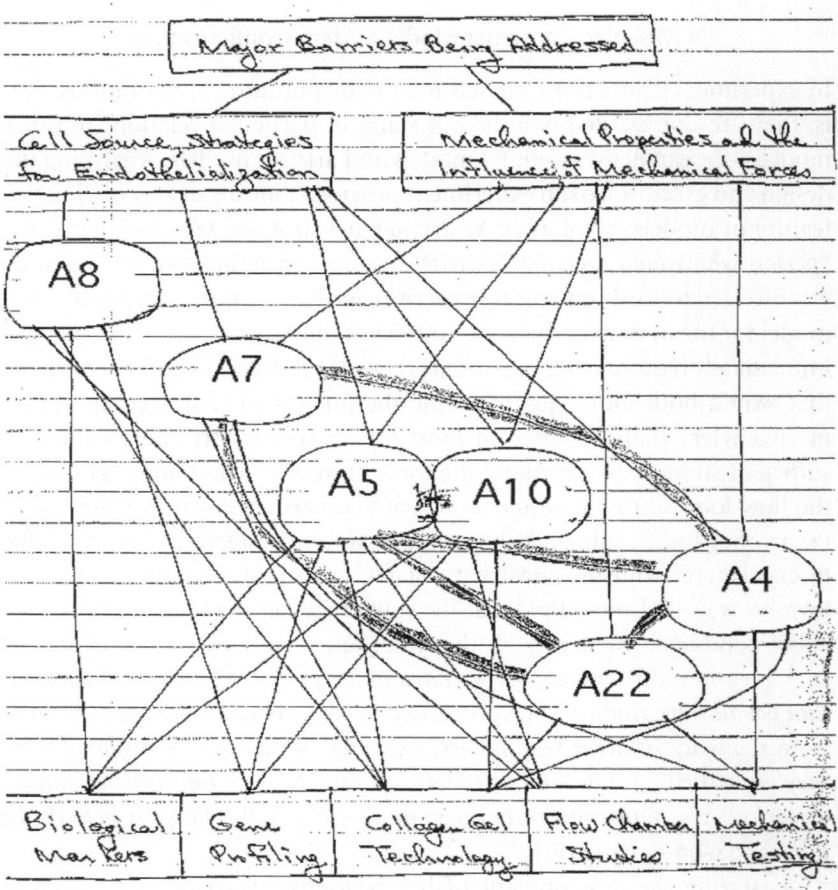

FIGURE 3.1. Diagram of the Lab A research space (drawn by A13).

("progenitor cells") derived from a patient's peripheral blood to line the vascular graft. The endothelial cells that line the artery are among the most immune-sensitive cells in the body. If the patient's own endothelial progenitor cells could be harvested and used, that would greatly enhance the potential of a vascular graft. However, the progenitor cells do not modulate thrombosis, which is a function of mature cells. A7 hypothesized that shear stress conditioning of the cell-seeded construct (by means of the flow loop) before implantation would solve the problem of platelet formation and the resulting thrombosis. Before turning to our analysis of her distributed model-based problem solving, it is instructive to examine her own succinct summary statement.

What is most notable from A7's account of her experiment is how she seamlessly interlocks in thought and expression biological and engineering concepts and models as she had enacted them in building and assembling the components of her model-systems and in conducting and drawing inferences through experiments:

> **We used the shunt to evaluate platelet deposition** *and that would be – in other words – were* **the cells, as a function of the treatment** *that they were given before they were seeded onto the* **engineered tissue,** *able to prevent blood clotting? And so we specifically measured the number of platelets that would sit down on the surface. More platelets equals a clot. So, it ended up being that we were able to look at the effects of* **shear stress preconditioning on the cells' ability to prevent platelets** *and found that it was actually necessary to shear precondition these blood derived cells at an* **arterial shear rate,** *which I used 15 dynes per square centimeter compared to a low shear rate, which in my case I used like 1 dyne per square centimeter, so, a pretty big difference. But I found that the* **arterial shear was necessary to enhance their expression of anticoagulant proteins** *and therefore prevent clotting. So in other words, the shear that they were exposed to before going into the shunt was critical in terms of magnitude, for sure* [2006–05–23-i-A-A7].

The boldface terms mark her reference both to interlocking conceptual models as they function in her understanding and reasoning and to inter-locked physical simulation models.

To make explicit the relation between her account and our model-system diagram, we unpack a few of her expressions. "*The cells*" are the endothelial progenitor cells she extracted from baboon blood and seeded onto the "engineered tissue" (vascular construct device). The "*treatment*" they received was "*shear stress preconditioning*" conducted by using the modified flow loop device. The objective of her research was to determine if and at what level the preconditioning ("*arterial shear*" simulation) of constructs would "*enhance their* [cells] *expression of anticoagulant proteins*" ("*prevent platelets*"). Through several iterations of the entire ex vivo model-system ("*used the shunt to evaluate platelet deposition*"), she found that the in vivo human arterial shear rate ("*15 dynes/cm²*") was required for sufficient protein expression ("*was critical in terms of magnitude*"). Likewise, instantiating her thought about the treatment of the cells by means of the model-system led to a revision of her conceptual understanding of the vascular construct so as to reflect the necessity of using that specific arterial shear rate to prevent thrombosis when using endothelial progenitor cells.

The diagram in Figure 3.2 provides a pared-down representation of our analysis of a vascular construct model-system specifically configured for this experiment. We created the diagram from our data quite a while before this final interview, but each component appears in her explanation. On our interpretation, it traces the construction, manipulation, and propagation of researcher and artifact models within the distributed problem-solving system that constitutes the experiment. In the figure, the models are highlighted by thick lines. The flow arrows represent the propagation of representations in the system as they are generated and manipulated.

To keep the diagram from becoming too complex, it does not include the numerous connections to other researchers in the lab research space (mapped in Fig. 3.1), but the construct, flow loop, and baboon models should be understood as communal achievements, representing years of research and collaboration. Each researcher (conceptual) model is a part individual, part community representation. Each device model represents and performs as some aspect of a cardiovascular system: The construct models represent and perform as selected aspects of the biological environment of a blood vessel, the flow loop model represents and performs as shear stresses on arterial walls, and the baboon model represents and performs as blood flow through a human artery. Neither artifact nor researcher models in this experimental set-up are fixed; rather they are representations that changed over time as each component was developed. There are three major components of this model-system: "construct" (left third of Fig. 3.2), "flow loop" (middle), and "baboon" (right third). It required more than 3 years of designing, building, and experimenting with the first two components before A7 could conduct the final ex vivo (animal) experiments over the next 2-year period. Model-system configurations, though continually evolving artifacts, are long-term investments. It is not possible to provide here a detailed account of the numerous cycles of designing, building, simulating, and evaluating the models represented in each part of the diagram, but we do provide a brief outline.

Just the making of the construct (left third of Fig. 3.2) required considerable novel research so that it could be used ex vivo. Issues of mechanical integrity (strength) became significant because the construct needed both to be able to withstand nearly normal blood forces and not to leak. The desired final experimental configuration required that this construct model be what the researchers considered the most "physiological" one the lab had created in that it instantiated all but the outer layer (adventitia), which A7 thought was not necessary to the experiment and might even grow on its own. First, endothelial progenitor cells (EPCs) were extracted from a

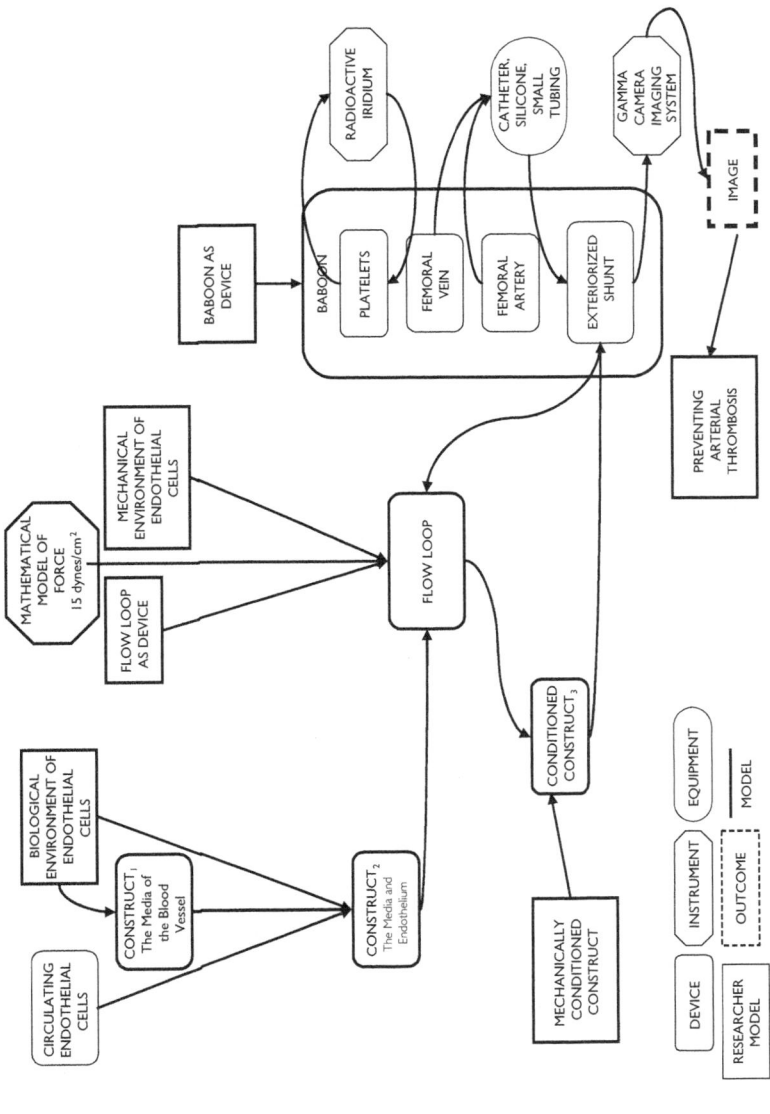

FIGURE 3.2. Partial vascular construct model-system.

67

baboon's peripheral blood. When working only with the EPCs, A7 used the
lab flow loop model at various shear rates and then used several instruments
to extract information about their protein and gene expression. Numerous
iterations of manipulating EPCs and using various instruments, such as the
flow cytometer to examine the cells for expression of molecules for thrombomodulin, facilitated her building a conceptual model of their function
and behavior in relation to shear stresses and contributed to building the
final construct device. Constructing this artifact model also required isolating an intact elastin scaffold, removing vascular smooth muscle cells from
the carotid arteries of baboon (sacrificed for experiments other than her
own), and determining the right collagen mix. Finally, because the construct
would need to remain in tubular form for both preconditioning with the
flow loop and for connecting to the baboon shunt, she needed to design
and build a new "cylindrical bioreactor" to encase it and thereby address the
problems of potential construct leakage, mechanical strength, seeding the
EPCs onto the tubular construct, and attaching the construct by suturing it
to the shunt.

To condition the constructs for implantation, A7 had originally proposed
redesigning the chamber of the flow loop to accommodate their tubular
shape. In the end a significant redesign turned out to be unnecessary. She
had the insight – through making an analogy directly with the shunt she
had designed earlier for the animal model – that it should be possible to
design an external shunt for the flow loop and attach the construct, now
encased in the cylindrical bioreactor she designed, to that shunt to simulate
a range of shear stresses (middle third of Fig. 3.2).

The animal model (right third of Fig. 3.2) was needed to provide a
functional evaluation of the EPCs in an environment that approximated
the in vivo environment more closely than would an artifact model. The
animal model comprises a baboon, surgically altered so that an exteriorized
shunt connects the femoral vein and the femoral artery, thereby allowing
a small amount of blood flow to be diverted through a construct during
an experiment. In the experimental set-up, a long tube is attached to the
shunt so that the construct can be placed directly in the gamma camera (a
commercially available instrument the size of a small table) to capture and
observe platelet formation during real-time blood flow. The baboon sits in
a specially designed restraining chair and has a suit over its body to prevent
it from pulling on the shunt. The baboon's blood is injected with iridium
so that platelets will be made visible by means of the gamma camera. The
construct is then disconnected from the baboon, and other instruments and
software programs are used to analyze the EPCs for information about the
expression of thrombomodulin, optical density, and electrical resistance.

Although the animal model provides a better physiological representation of in vivo human phenomena, it is less reliable than a fully in vitro engineered system because *"in the lab we can control their [endothelial cells] environment completely.... We can control exactly what flow is like and we can monitor by visually seeing it. But when we move to an animal model it's more physiologic – the challenge then is that it's a much more complex system"* [2004–11–29-i-A-A7]. To advance the research, some control and precision needed to be sacrificed so as to better exemplify complex characteristics of human physiology. A7 made several trips over 2 years to the out-of-state lab where the animal models reside to run experiments and modify her simulations. In the end, she was able to prevent platelet formation. Within the framework of distributed cognition, it is the entire *system of researcher-artifact-animal models* through which the inference was made that arterial shear is necessary to enhance the expression of anticoagulant proteins. This inference then led to the hypothesis that, if human EPCs were to be conditioned at the normal arterial sheer rate before vascular graft implantation, they would function as mature endothelial cells with respect to modulating thrombosis.

PROBLEM SOLVING IN A NEURAL ENGINEERING LABORATORY

The assumption that guides research in Lab D is that advancing understanding of the mechanisms of learning requires investigating the network properties of neurons. Historically, studies of living neurons have been conducted by single-cell electrophysiologists who seek to understand the fundamental properties of a single neuron, whereas network studies were done on computational models of neural networks. Early attempts at investigating the nature of living networks of neurons depended on cutting slices of a brain, fixing the neurons (in formaldehyde), and then looking at the physical structure of the neurons underneath a microscope. The Lab D director was not happy with looking at dead neurons and instead wanted *"to look at living things, while the interesting parts would happen."* In the early 1990s he began formulating the goals that drive the research in Lab D (founded in 2002). As he stated,

> But then I thought, the idea hit me – I can't exactly say where it came from – which was perhaps you could make a cell culture [model] system that could learn something. And I thought to do that you would have to have the capability to see the cells while they were doing their learning.... And so I thought, OK, well I figured, if we were gonna create an in vitro learning system, it has to be somehow connected with the outside world. In order

*for learning to have any definition at all, there has to be some way for it to
behave and to see whether there was some change in behavior there had to
be sensory input to see whether you could even influence that. Those were
my prerequisites for learning* [2002–12–11-i-D-D6].

To further this goal of creating an in vitro learning system, he spent his
postdoctoral years designing and perfecting what has become Lab D's most
prominent model-system, "the dish," which we discuss later, and also developing the technology for imaging it. With these in hand he established his
own lab in which researchers see themselves as breaking ground in "*mezzo-
level*" neuroscience. That is, they investigate the nature of networks of
cortical neurons, which are too small and unorganized to be considered a
functioning brain, but are at a higher level of organization than the single
neuron.

   Although the main efforts at problem solving are directed toward developing a basic understanding of neural mechanisms, extending the research
to applications in medicine and engineering was part of the framing of the
lab and part of its current activities. In the initial interview just quoted from,
the lab director also explained the "*biggest picture goal*"; this explanation
provides a glimpse at the wide open horizon in which the lab situates its
research agenda:

   *A problem that gets me out of bed in the morning – and motivates what I
   want to do – is that we're, I don't think that human beings are smart enough
   to deal with the intelligence they have. So, if we were a lot stupider, we'd be
   fine; if a lot smarter, we'd be fine. We are at some intermediary level, being
   smart enough to screw things up but not smart enough to deal with that.
   So the big picture problem that I hope to at least make some progress on is
   how to make human beings smarter. And that might be way in the future –
   the payoff that I am anticipating, but first we have to learn what human
   intelligence is. . . . So, in a nutshell I suppose everything I'm doing has some
   connection with that* [2002–12–11-i-D-D6].

As we did with Lab A, in the next sections we discuss some of the central
components of model-systems in Lab D.

                               The Dish

Simply put, "the dish" is a network of cortical neurons living, not as part
of a larger brain, but as a small network in a Petri dish. Embedded in the
bottom of this Petri dish are 64 electrodes capable of recording and injecting

electrical activity in the network. The basic components of the dish are as follows:

- *The MEA*: The microelectrode array (MEA) is a small, glass, Petri-style dish that has an 8 × 8 grid of micro-electrodes embedded in the bottom of it. The electrodes poke upward from the bottom of the dish into the neurons.
- *Cells*: Lab D uses parts of a rat's cortex as their network. The two basic types of neural cells, neurons and glia (support cells), are harvested from the rat and then cultured in this dish.
- *Medium*: The lab uses a sugary cocktail of biological chemicals to feed the cells.
- *The lid*: A thin film made out of Teflon is stretched over the MEA and held in place by a thick O-shaped piece of Teflon (like a slice of pipe). Teflon is used specifically because it is nontoxic and allows O_2 and CO_2 through so that the cells can breathe. However, it is not porous, so it keeps everything else (such as bacteria or fungus) in or out, even when it is stretched.

The design and construction of the dish of cultured neurons interlock concepts, methods, and materials and embed constraints from biology, chemistry, and electrical engineering. The dish is a model-system itself because it incorporates both engineering components and biological materials. The primary focus of the researchers' problem-solving activities while we were conducting our investigation was to understand the nature of this novel model-system. Understanding the dish not only frames the inquiry but also plays an integral role in their overall research progress. Only after the researchers understand their physical model at its current level of approximation can they progress in the construction of new models that more closely parallel in vivo situations.

Construction of the dish has involved numerous technical problems. To overcome some of these, Lab D has chosen to simplify the model to a single layer of neurons to reduce the number of possible variables in the system. Doing so provides a more restricted set of questions to ask in developing their understanding of network processing. However, they believe that this simplified model provides a close enough approximation to yield valid information, although determining what is necessary for the dish to be a model is an ongoing research problem. As the lab director said,

> So there are a lot of people out there who are doing very detailed modeling
> of single neurons, but I'm much more interested in what happens when you

take a bunch of cells and have them interact with each other, what sort of emergent properties pop up. So, it probably isn't necessary to include all the details that you would find in a [single neuron] model, with ions and stuff, but it may be, so that's part of our job is to find out which of the details of biology are important in these sort of network properties and network phenomena and which are sort of incidental [2002–12-D-D6-tour].[1]

In constructing the dish model-system, the neurons are disassociated (their connections are broken apart to the point that they are free from each other), plated (processed and placed on the MEA), and then cultured so that connections are allowed to grow. D4 describes the reasoning for using this procedure to construct the dish rather than plating a brain slice on the MEA:

Yeah, one single-layer of neurons. We try to get them down in a monolayer. That's the whole idea. It's a simpler system to study than –. You could do slices that are not disassociated cultures but the problem in slices is it is difficult to maintain them. You have to kind of get the fluid [in] *right. Sometimes only the outer cells – you know the medium does not go into the inner layers. The inner layer dies off but the outer layers are fine, stuff like that. So, it's more difficult to keep them alive. And we want to study over a long term, so we want to keep them alive over months, years* [2003–03–20–i-D-D4].

Generally, the researchers view the dish as an in vitro model of basic in vivo neurological processes. However, there is some contention among the lab members over what it models more specifically. Some maintain that the dish is "*a model of cortical columns*," whereas others think it "*is a model of development* [of the brain]"; when pressed, some even admit that "*it may just be a model of itself.*" However, they all agree that studying the dish will yield an understanding of the basic workings of network-level cortical neural processing, as D4 explained:

First of all, it's a simplified model. I say that the model is not – it's artificial, it's not how it is in the brain, but I still think that the model would answer some basic questions, because the way the neurons interact should be the same whether it's inside or outside the brain. . . . You know, because we are in an artificial environment, it's not the same, you know, it's not the same concentration as it's in the brain; nothing is the same. I'm growing in an external environment. But, I think that the same rules apply [2003–04–09-i-D-D4].

---

[1] Quotations marked with "tour" are not from the interviews, but from a tour of Lab D conducted by the lab director for us and some other guests. We did ask questions during the tour.

Thus, the dish is best understood as a generic model of cortical processing behavior and function. It provides a physical simulation of how networks of neurons – in general – communicate and process information. This generic understanding of the networks is not the end goal, however. On this generic understanding, the lab members plan to build a more refined understanding of neural networking processes:

> *Clearly it's missing a lot of other brain parts that are important in what brains do. I'm assuming they are important. And at some point we might be studying cultures with different brains parts mixed together, or specific three-dimensional pieces that are put together* [2002–12-D-D6-tour].

Although all of the researchers use the dish, the researchers pursue several different avenues in this lab, including pharmacological studies, morphological (imaging) studies, and simulation with both computational models and physical models. Here we focus only on the lab's electrophysiological investigations in which the central problems have been to devise ways of communicating electrically with the network of neurons (what members call "*talking to the dish*") and then, in turn, trying to understand what it is "*saying*" back to them in response, as they attempt to produce learning in the network. The core of the method involves electrically stimulating the neurons and recording the electrical response of the dish. Generally, this part of the research involves learning to interpret, control, and mathematically represent the dish's behavior under various conditions of stimulation.

The main problem driving the experimentation in Lab D's electrophysiology research is to understand how information is encoded and processed in the dish networks. It would be ideal to have direct readings from every single neuron in the dish. However, this is impossible with current technology. Consequently, access to the dish is mediated by a comparatively small set of electrodes (the MEA). The signals ultimately received are a representation of neural activity filtered through several models that are discussed in this section. Instead of studying the full neural activity in the dish, researchers study the "spike" data recorded from the electrodes of the MEA.

The notion of a spike derives from the single-cell experimental paradigm and refers to the electrical trace left behind when an individual neuron fires. In single-cell electrophysiology, it is possible to read electrical activity directly from the neuron, and as a consequence, the model for neural firing is well known: There is a steep jump in voltage potential as the neuron depolarizes (fires) and then a proportional drop in potential as the neuron recovers. Researchers pursuing multicellular neuroscience have borrowed the notion of a spike, but have modified it to suit their situation. The researchers of Lab D estimate that the electrical activity recorded on a single electrode can come from, on average, three to five different neurons. When dealing with the electrical traces of many neurons, it is possible (and is often the case) that several neurons around a single electrode will fire simultaneously. As a result, a spike can represent the firing of more than one neuron. However, because the neurons are all on one trace, it is impossible to tell the difference between the firing of a single neuron and the firings of multiple neurons.

Historically, spikes were tagged by hand. However, Lab D has automated the process of identifying spikes by developing a piece of software its researchers call the "spike detector." The spike detector embodies the lab's model of a spike, which includes the spike's "height" (difference from the average voltage) relative to the noise on the electrode and the "width" (duration) of the change, along with a few other more subtle characteristics. This software checks for jumps in the voltage that match this model of what a spike should look like, tags the spike, and keeps a snapshot of the electrical activity immediately around the spike. The researchers begin their analysis with the filtered spike data.

Several other pieces of lab-developed software are used during electrophysiological experiments. These are collectively referred to as "filters" and perform a number of different transformations on the raw neural data before the information ever reaches the researcher. Each of these filters embodies a model of an aspect of the neural data. However, it is possible that the filters miss actual neural firings or detect a jump in the readings that does not correspond to any action potential (i.e., provide a false positive). Thus the meaning of an individual spike needs to be understood in terms of the filter algorithms that created it. The researchers are intimately aware of these processes and perform their analyses in light of this understanding.

In the simplest form of analysis, the spike data are transformed into a visualization rendered on a computer screen. The lab uses a number of different types of visualizations for these basic data and far more for their higher level analytical transformations. The simplest visualizations come from the data-capture software they have developed called "MEABench." As the computer is capturing neural data, MEABench displays the neural activity in real time.

In sum, the researchers' ability to "*see*" the electrical activity of the dish involves a number of interconnected mental, physical, and algorithmic models. The visualization is a representation of the information produced by the filters. The filters are instantiated models of Lab D's understanding about neural signals. The signals themselves are a model of neural communication (they abstract away other factors such as chemical signaling). Finally, the dish itself is a model of cortical neural processes. This series of interlocking models is the base model-system for any electrophysiological experiment run within Lab D.

However, this particular set-up is the most basic model-system used in the lab. It only enables open-loop experiments; that is, experiments in which there is no feedback. Yet in vivo learning in animals and humans involves feedback to the brain from embodied experience in the world. To address the problem of embodying the dish, the lab has developed two kinds of closed-loop families of model-systems: the "animat" and the "hybrot" model-systems. Hybrots comprise various kinds of robotic devices connected to the dish. Animats comprise computationally simulated animal worlds that are connected to the dish. Here we look at one animat model-system configuration that formed a major component of the research on embodied learning conducted initially by D2 and later by D2 and D11 together.

An Animat Model-System Configuration

[In] the traditional way to do in vitro physiology . . . the closest thing to behavior is little waves on the oscilloscope screen. [It] has nothing to do with any behavior, other than light on the screen there. And there is not any sensory input other than electrical pulses through a couple of electrodes. You know, it is very disconnected, and one of the things that was really shaping my thinking at that time was this book here. This is the first of a proceedings from this conference – simulation of adaptive behavior, here. And, I think it was 1990 that they had this conference, yeah. [A]ll of the people in that book are simulating animals or what they call "animats." The term was coined around that time by the guys who were at this conference. They were simulating these things on the computer or they were building robots that were animal simulations. They were continually reemphasizing the importance of embodiment and being situated [2002–12–11-i-D-D6].

Originally, then, the idea for animat model-systems came from the domain of computational modeling. There the goal was to create a very simple model "world" and a very simple model "animal" and then simulate the activity of the "animal" in the "world." The lab director borrowed this idea of modeling animals, but decided that he could improve on it. Whereas others were using a completely computationally simulated animal, he set out to make a partly living model of an animal by using the dish as the "brain." The term "animat" in Lab D, then, refers to a computationally simulated entity that is controlled by the activity in a dish. An animat model-system consists of the dish (the "brain" of the computational animal) and two translation programs: one designed to be the simulated sensory input apparatus and one to simulate motor output that exists in a simulated environment. In short, animats are used to simulate the embodiment of the neural networks and as a rudimentary model of an animal functioning in the world.

In the dish experimental set-up described in the previous section, the electrical signals produced by the neurons are simply recorded and then analyzed. In contrast, animats are part of closed-loop electrophysiology. Closing the loop simulates embodied learning. The electrical signals produced by the neurons are not simply recorded, but instead are transformed so as to provide "sensory information" in a way that is meaningful in the context of the experiment. This sensory information is run through a translation program that converts it into a pattern of stimulation to be administered by a stimulation board. The stimulation produces an electrical response in the dish that is recorded and run through a separate translation program that converts the signals into "motor" commands for the animat. The motor commands move the animat in its "world," and this movement, in turn, produces a change in the "sensory" information. The change in sensory information is again read from the electrodes in the dish, and the loop continues.

As expressed by the researcher, the animats provide model-systems for studying the relationship between the fundamental nature of information processing in the neurons and the visible behavior of animals:

> There are lots of different arguments for it. I think probably one of the best arguments is – if you look at what neurons do they – they learn things. And that's what we're trying to figure out . . . how learning works, how memory works. And so if we have neurons in a dish, we want to see them learn something – make associations – it's a bit more obvious to see the learning involved in a closed loop [situation].' Cause you define what it is – it's going to learn based on – the body you give it and the environment you allow it to work in [2004–10–14-i-D-D2].

An animat model-system serves to demonstrate network plasticity more convincingly than open-loop experiments and should lead to a better understanding of how to interpret and control the activity in the dish. The diagram in Figure 3.3 provides a pared-down representation of our analysis of the generic animat model-system. Just as there are many different types of creatures in the real world, Lab D has created an entire family of model

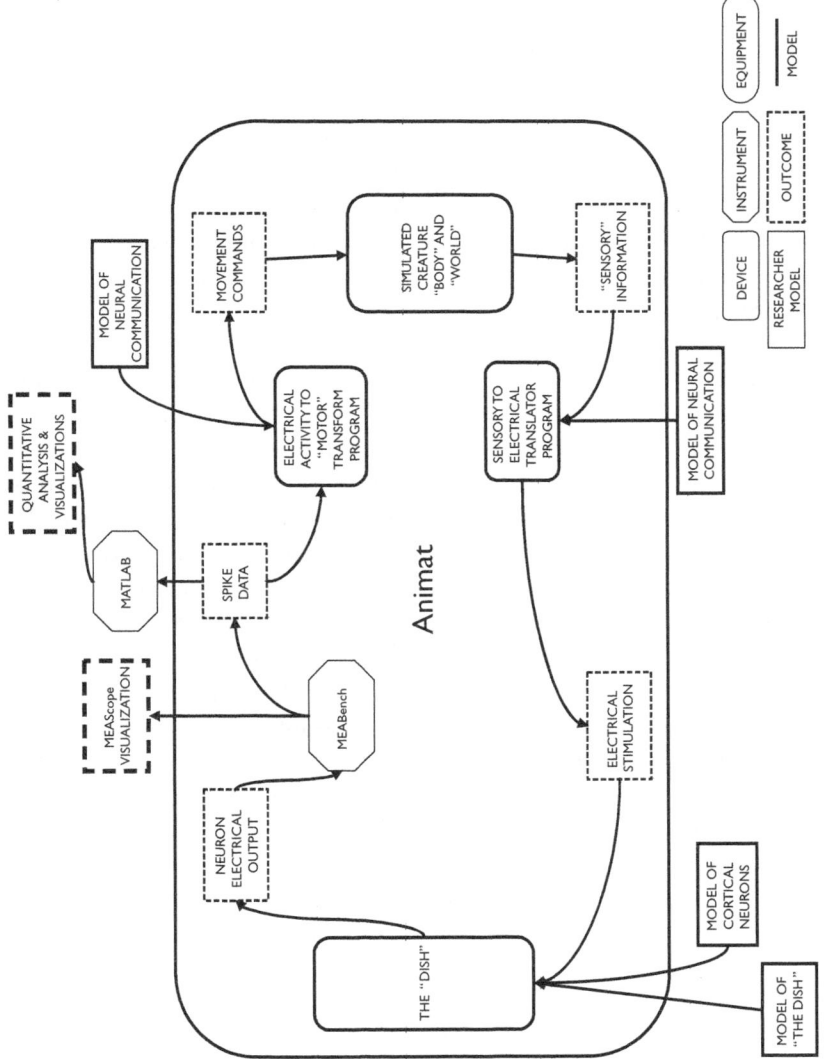

FIGURE 3.3. Partial animat model-system.

animats, one of which is a simulated "moth." Here we briefly consider D2's construction and use of the moth model-system:

> *The original model is basically, I have a circle and then the center of the circle, which this is the environment. The center of the circle, where you can find what you can think of as a light . . . and I wanted, the moth, which would be the animat, would move towards the light like moths do* [2004–11–15-i-D-D2].

In the case we analyzed, the screen visualization comprises a circle that delimits the entirety of the moth's world, a dot in the center represents the simulated "light," and lines across the circle represent the paths that the moth has followed during the duration of the experiment. The moth was given a simulated "eye" to "see" the light and the simulated motor ability to move around in the "world."

The two most interesting parts of any animat experiment are the two programs that interpret the neural signals. The sensory translation program enables the moth's sensory system to "see" where the light was and "see" where it is itself, and it turns this information into a series of neural signals just like an animal sensory system. Once the dish is stimulated by this sensory input, it responds with its own electrical activity. This activity is recorded and processed by the method described in the discussion of electrophysiology, and then it is translated into motor commands for the moth. Here the neural activity can be treated as a population vector; in other words, each electrode is taken to represent a possible direction of motion. Activity on the left of the dish would indicate that the moth wanted to move left, activity on the right would mean the moth wanted to move right, and so on. In essence, the network as a whole determines which direction to go in, based on the sensory input. Importantly, the translator programs arise from a number of different models. First and foremost, the programs stem from a model of how D2 understands neural communication, and so the translation algorithms instantiate his idea about how neurons communicate. Taken together, the translator programs form a generic representation of how an animal's sensorimotor system operates.

In sum, an animat model-system, such as the moth, is an in vitro simulation of in vivo embodied learning. The sensory translation program derives from the researcher's understanding of how animal sensory systems translate raw sensory information into meaningful information for the cortex to process. The dish component is a model-system that represents basic neural processing. The motor translation program embodies a model of how an animal's motor system converts neural output into motor function. The output from this program feeds back "motor" commands to the animat's "brain," the dish. Taking the artifact model-system together as a whole, the experimental animat is a highly simplified model of learning in a real-world animal.

As with most model-systems in these types of cutting-edge communities, the animat system is constantly evolving. Using the results of the first moth, the researcher revised his understanding of neural processing and, in turn, revised his model of moth. After running the moth a number of times, he updated the sensory control system; in particular, he *"simplified that even further just to have frequencies only dependent on positive X production.... Just to make it, easier to analyze in data analysis... Just to try and simplify everything. I could just say, 'Okay, this part of the stimulation does this'"* [2004–11–15-i-D-D4]. Much like the lab's decision to use monolayer cultures, here the researcher chose to design a simplified, more easily understandable model. This abstract representation of an animal gives the researchers less degrees of freedom, but enables the development of more clearly defined questions and experiments that provide results that are less complex and more easily interpreted.

With these descriptions of the central model-systems that participate in the problem-solving processes of Labs A and D in hand, we now turn to a discussion of the problem-solving person as enacting these practices within distributed cognitive-cultural systems.

ENACTING DISTRIBUTED PROBLEM SOLVING

In this final section of the chapter we discuss three interrelated categories emerging from our coding – cognitive partnering, model-based cognition, and interlocking models. These categories are particularly apt for conveying the idea that problem solving is performed by means of a distributed cognitive-cultural system comprising people and artifacts.

Cognitive Partnering

In these laboratories, problem solving requires forming relations with others and with select artifacts – those that perform model-based simulations. Thus, we understand "cognitive partnering" as an expression of cooperative participation within an epistemic culture that enables or sustains particular cognitive-cultural practices. There are two overlapping varieties of cognitive partnering: partnering among persons, for which we have distinguished two subcategories – peer-to-peer partnering and mentor-to-apprentice partnering – and person-to-artifact partnering.

Person-to-Person Cognitive Partnering
An interview passage from Lab A illustrates how we understand person-to-person cognitive partnering of the peer type. New to the lab, A22, a master-level student, discusses how she understands the research in which she will engage:

> I: What's your notion or sense of where your research is headed?
> A22: *I'm going to be working with – I still need to learn more about it but I'll be working with the fabrication of constructs with the possibility of either looking at different methods of fabricating constructs, or making basically a construct within a construct.*
> I: So what – I've never heard that before – "construct within a construct?"
> A22: *Yeah, this was an idea that A13 had. I guess A11 has actually done something similar, but looking at doing a construct with an inner layer that the collagen with smooth muscle cell and then an external layer which is a collagen, which is simply collagen without muscle – without cells – and then, and using that. I guess A11 actually did something that was very similar but in reverse order, with the smooth muscle cells on the outside* [2002–10–10-i-A-A22].

The second agenda option A22 gives here, that of making "*the construct within a construct*," is most interesting because she frames it as an idea originated by another lab member, A13. Although she is unclear as to whether the idea or method originated with A13 or A11, she is clearly aware that her research will develop in relation to that of other lab members. In either case, both the notion of "*the construct within a construct*" and the practice of building such a device are specific to the culture of Lab A. As such they are identified initially with a particular member of Lab A and credited to him (A13). The making of a "*construct within a construct*" is distinguished in this sense from activities such as cell culturing that would be common to all tissue engineering labs. A22 seems to have appropriated the notion and the practice from another lab member that affords ways for her to frame her own research agenda (and answer the question concerning where her research is heading). Moreover, the construct, a simulation model, immediately enters her account as central to her research. A22's learning is revisited in depth in Chapter 8.

A7 provides an instance of cognitive partnering engaged in by a more mature researcher. As we saw in Section 2 and as illustrated in A13's diagram of the research space (Fig. 3.1), A7 engaged in cognitive partnering with several members who were working on aspects of the design of the construct and flow loop model-system that she would use in the problem solving we described. Further, she provides an example of how researchers partner with researchers outside of their lab:

> I: Where did you get the idea of how to use the shunt [with the baboon], and is this a set-up that you had to develop?
>
> A7: *No, definitely not in terms of me developing it. I think it's really, really cool, but X whose lab you know I went to visit. . . . I don't think he was actually the first . . . he definitely worked with Y, and I think the group at [institution] has also used primate shunts before. But he's probably the person who's used it most extensively over time. So, it's kind of a – it makes for a perfect collaboration, and Z, who is a scientist who works with him, is the person that I work with directly you know, in the [other] lab everyday, and has the expertise to kind of help me [laughs] and we have something we want to investigate, so . . .*

The cooperation facilitating cognitive partnering need not be purposeful or explicit, though it is more so in some cases than in others. Within the same interview, for example, A22 describes several forms or styles of cooperation facilitating her learning of cell-culturing techniques (mentor-to-apprentice partnering):

> A22: *I learned some from A5, and I've worked with A7, and I'm probably going to work a little bit with A10 this morning too.*
>
> I: And how do they – how do they teach you to do that?
>
> A22: *Well, A5 – we sat down and she worked with me. She showed me: we got cells, we had six different groups, she showed me like three of the cell cultures, and then she watched me and corrected me as I did the other three. And then I just came back and kept doing it and if I had any questions I could ask. And with A7, mostly I've just been watching her work, so far. And then today, this morning I'm going to be doing trypsynizing where basically A10's going to tell me what I need to do and I'm just going to watch. I mean he's going to watch me and make sure that I do it correctly* [2002–10–10-i-A-A22].

Thus, A22 is experiencing several partnerings that enable her to acquire the cognitive practices surrounding the development and use of the central artifacts of the research in this lab. What it means for learners to form cognitive partnerships with other researchers shares features with the notion of "cognitive apprenticeship" (Brown, Collins, & Duguid 1989) and is fairly straightforward. However, we developed the notion in the first instance to capture a shift in the way learners think and talk about and interact with their devices as they progress in their research.

Person-to-Artifact Cognitive Partnering
Cognitive partnering with artifacts is an expression of working cooperatively with specific artifacts to carry out particular research goals. An explicit example comes from Lab D:

> *Well, that way I'm strengthening that particular pathway, so the network would prefer to always excite these two cells in a certain way after my modifying input* [dish model]. *Maybe they weren't before my modifying input, and after my modifying input the pathway becomes stronger* [2003–04–09-D-i-D4].

Our understanding of cognitive partnering between person and artifact has been informed in part by what we observed as a frequent practice in the researchers' description of their work – namely the attribution of something like *agency* to objects, artifacts, and devices – which is most clear in the practice of anthropomorphizing. Significantly, the language used to discuss the devices shifts from that of the newcomer who discusses "*that thing over there that has hooks and pulls*" to that of the advancing researcher who increasingly uses anthropomorphic language as he or she becomes more expert. Although much of it concerns cells, which are living entities, importantly, this language use extends to nonliving parts of model-systems as well, as seen in these examples:

- "*The cells once they are in the construct will reorganize it and secrete a new matrix and kind of remodel the matrix into what they think is most appropriate*" (construct model – Lab A).
- "*[The board] sees voltages in different way*"s (bioreactor model – Lab A).
- "*Yeah, seven parameters it has to look at in order to decide what's a burst*" (burst detector software in dish model-system – Lab D).

We construe such anthropomorphizing as signaling the researcher's *attribution of agency* to the artifact. As the researcher matures, the simulation device comes to be conceived as a "partner" in research. In one sense, cognitive partnering marks the researcher's coming to understand the research through the lens of what the device affords and constrains. Yet this understanding goes beyond that to conceptualizing the devices as possessing quasi-independence – as distinct from and interacting with the researcher.

In the course of experimenting, the artifact reacts and interacts often in undesired or unexpected ways, which leads the researcher to come to treat the device as an active participant in the investigative process.[2] Finally, this language signals too that there is an affective dimension of research, as there is in relationship formation generally. For instance, this dimension is evidenced by a researcher in Lab A telling a newcomer to *"think of them [cells] as pets"* and by the frustration of a researcher in Lab D with the behavior of her device: *"Pfft – you keep them [the dish neurons] happy by feeding them, by taking care of them, hopefully stimulating them,* [in a motherly and condescending voice] *and telling them to do something! I don't know what to do to make them happy. I don't know – make them happy!"* If the MEA dish is not happy, neither is the researcher. We explore this dimension further in Chapter 4.

We construe such expressions as indicating a sympathetic engagement with artifacts and devices and a tacit sense of working together with them toward problem-solving goals. As noted, this practice becomes more pronounced as members progress as researchers – as they develop a deeper sense of research as a cooperative enterprise not only with others but also with the devices essential to carrying out their research. Moreover, the notion of partnership used to express relations in laboratory cultures seems fitting in view of the affective, motivational, and identity aspects of research practice that we explore in later chapters. Affective experiences of researchers in relation to one another and to their work, including transformations in their feelings of belonging and investment in the work of the laboratory, cannot be eliminated from the cognitive-cultural system that creates and sustains problem-solving practices.

Distributed Model-Based Cognition: "Putting a Thought into the Bench Top"

Solving problems in the cognitive-cultural research space of a laboratory, we claim, requires specific kinds of partnering of persons and of persons and

---

[2] Our notion of cognitive partnering might appear similar to the "actor network" theory introduced by Bruno Latour (1987). However, there are two salient differences: First, on our account not all "actors" are equal (i.e., there would be no distributed *cognitive* system without the human agents), and second, "agency" as we understand it implies "intentionality," so the artifacts can perform cognitive functions in the system and exhibit independent behaviors and activities, but are not themselves agents. The point we are making is that learning and research require researchers to form relationships, and one type that we witnessed is a relationship with the cells and simulation devices, which involves researchers attributing agency to these artifacts, living and nonliving.

artifacts to achieve cognitive goals. To highlight specifically the problem-solving practices enabled by this partnering, we now examine and further develop what we have elsewhere construed as the "coupling" of internal (mental) and external (artifactual) representations in problem-solving processes (Nersessian 2008a; Osbeck & Nersessian 2006). The notion of "coupling" provides a counterpart to that of partnering. One *has* a partner, but one is only a *part* of a couple. The argument presented here is that what the Lab A researcher referred to as the *"experimental model that predicts"* comprises the entire model-system: researcher(s) mental models and artifact models as represented in the diagrams of Figures 3.2 and 3.3.

Distributed cognition seeks to reconceptualize cognitive processes as comprising systems of humans and artifacts. To date distributed cognition researchers have focused much of their attention on the artifacts that participate in cognitive processes and on the roles of the humans in coordinating the generation, manipulation, and propagation of artifact representations as they accomplish cognitive work. Of course, as Hutchins and others have noted, brains are parts of cognitive systems, and perceptual and motor processes play a role in coordinating artifact representations (Alac & Hutchins, 2004; Becvar, Hollan, & Hutchins, 2008). However, the nature of the representations that the human components contribute to distributed processes has received scant attention. We propose extending the traditional notions of "mental model" and "mental modeling" by means of the notion of "distributed model-based cognition," which comprises the researcher and artifacts models. Our data provide evidence that researcher representations are themselves model-like in structure, possessing simulation capabilities when used in reasoning processes. On our interpretation, the device model-system and the researcher mental models constitute a distributed inferential system through which problem solutions are achieved by means of simulation processes. We understand this to mean that part of the simulation process occurs in the mind or imagination and part in the real-world manipulation of the model. This is what we mean by calling the mind–artifact system a coupled inferential system. Inferences are not achieved by the scientist(s) alone, but by the scientist(s)–artifact model combinations.

The interview data provide evidence that, when researchers actively think about and reflect on the in vivo phenomena and the in vitro devices, they are thinking by means of mental representations that have a model-like structure. That is, the mental representations are organized representations of the parts of physical systems and interactions among them, which are customarily referred to as "mental models." The data on how researchers explain, predict, and otherwise reason support our interpretation that the long-term conceptual representations they draw on also consist of organized understandings comprising representations of the structures, functions, and/or behaviors of the phenomena they are investigating and of the artifact models they create for experimenting.

The mental models literature comprises a wide range of cognitive science research that posits models as organized units of mental representation that are used in various cognitive processes, including reasoning, problem solving, discourse comprehension, and sense-making. As Nersessian has noted in previous writings (Nersessian, 2002, 2008a, 2008b), there are two usages of the term "mental model" in the literature: One addresses the nature of long-term memory representations (see, e.g., Gentner & Stevens, 1983), and the other addresses the nature of working memory representations in inferential processes; that is, the activity of "mental modeling" in discourse interpretation (see, e.g., Perrig & Kintsch, 1985), logical reasoning (see, e.g., Johnson-Laird, 1983), and mental simulation (see, e.g., Nersessian, 1992b, 2008a). In addressing the question of how engineering scientists think, we invoke the notion of mental models to specify the contributions of the researcher components of the distributed problem-solving systems. Our data support the hypotheses that researchers both represent physical systems by means of long-term memory models and use model representations in working memory inferential processes. In science, and outside of it, one means of reconceptualizing mental modeling is to describe it as a form of practice, thereby placing the emphasis on represent*ing* rather than representation, as a sense-making activity of persons (in relationship with other persons and with environments). The products of this activity (representations) constitute dynamic interpretations that afford new opportunities for problem solving and creative rearrangement through manipulation (Osbeck & Good, 2005).

How do researcher mental models and external model-systems function together as components of distributed cognitive processes? Thinking about this question from the science end of the problem-solving continuum has led us to extend an earlier hypothesis that Nersessian formulated for conceptual simulations to the distributed simulations considered here: A significant basis of the human component's participation is the capacity for simulative mental modeling (Nersessian, 2009). The earlier hypothesis she advanced was that "in certain problem-solving tasks humans reason by constructing a mental model of the situation, events, and processes in working memory that in dynamic cases can be manipulated through simulation" (Nersessian, 2002, p. 143).[3] This hypothesis derives from the original conception of

---

[3] Significant cognitive science research supports the hypothesis that simulation is a fundamental form of computation in the human brain (see Barsalou, 2008, for an overview).

Kenneth Craik (1943) and extensive experimental literatures over the last 25 years of research on mental modeling, mental animation, mental spatial simulation, and embodied mental representation (Nersessian, 2008a). The original hypothesis arose from examining a range of historical records of scientists' reasoning. Recent psychological research, based on observational studies of scientists and engineers as they try to solve research problems, lends support to the hypothesis and further details the nature and role of conceptual simulations.[4]

To this point we have been using the expression "mental models" to connect our analysis of scientific problem solving with the extensive literature on mundane cognition. Before proceeding to extend that notion to distributed problem-solving processes, we want to shift to a more neutral descriptor. The traditional correlates "mind–body," "mental–physical," and "internal–external" (to the body) are laden with dualistic assumptions (Nersessian, 2005). Therefore, for the rest of the analysis we use the terms "researcher model" and "artifact model," where the former refers to model representations and processes typically understood to be related to the human body and the latter refers to other model components of the distributed cognitive system. We now use our synthetic view of mental models to bridge traditional and distributed views of cognition by considering how what is usually called "mental modeling" – a dynamic inferential process – can be understood as a distributed cognitive process in which researcher and artifact representations are interlocked. Within the distributed cognitive system, inferences arise from processing information in the researcher's memory and in artifact model components.[5]

One researcher characterized what they do in experimenting by means of simulation models as *"putting a thought into the bench top and seeing whether it works or not."* We found this characterization particularly apt. As an instantiated thought, the device is a physically realized representation with correspondences to the researcher model. As a tangible artifact, it and its interpretation evolve along with the researcher's understanding in experimental processes. As a representation, it refers both to the researcher model and to the in vivo phenomena. Simulations with artifact model-systems are integral to the researcher models in that they are intended to

---

[4] See, for example Christensen and Schunn (2008); Trafton, Trickett, and Mintz (2005); and Trickett and Trafton (2007).

[5] For related views of mental modeling see Gorman (1997); Gorman and Carlson (1990); and Greeno (1989). Further, the evolutionary psychologist, Merlin Donald (1991) has argued that evolutionary considerations lead to a distributed notion of human memory as a symbiosis of internal and external representations.

function epistemically. In the search for knowledge about in vivo systems, researcher and artifact models can interlock both in concurrent simulation processes, in which building and simulation activities can lead directly to transformations to the researcher model; they can also interlock over the longer term, through processes that bring about changes in long-term memory representations. Construed in this way, the artifact model is a site of simulation not just of some biological or mechanical process but also of the researcher's understanding.

A simulation with an artifact model can be viewed as an extension of the scientific practice of thought experimenting (what a Lab D researcher called "science fiction"), though one that is more complete and complex and potentially less subject to individual bias and error. It affords a more detailed representation and a wider range of manipulations than a thought experiment. In a manner similar to how microscopes and telescopes extend scientists' capacity to see, artifact simulation models extend their cognitive capacity for simulative reasoning. As with a thought experiment, the model-system simulation provides predictions about how the in vivo system *might* perform under specified conditions. Whether one is warranted in transferring these to in vivo phenomena depends on the kinds of considerations discussed briefly in the sections on modeling practices in each lab, which concern the fit between the representations and manipulations performed and the in vivo phenomena. For instance, is a first-order approximation sufficient for representing blood flow? Do the effects of shear stress on the cells in a flat construct differ in relevant ways from the effects on tubular arteries? And so forth.

As with thought experiments, simulations with models provide a kind of counterfactual reasoning. Recall A7's explanation of her vascular construct model-system: *"This is an experimental model that predicts what would happen – or you hope that it would predict – what would happen in real life"* [2006–05–23-i-A-A7].

The argument we are making is that the *"experimental model that predicts"* is the distributed model-system comprising researcher and artifact models. In experimental processes, coupled researcher and artifact models form distributed experimental configurations from which inferences flow. Reasoning by means of this system involves co-processing information in human memory and in the environment. The nature of these processes and the mechanisms connecting brain and artifact components are important open questions for psychology and neuroscience. Research needs to address both cases of concurrent, interactive manipulations of working memory researcher models and artifact models ("coupling") and cases in which simulation is carried out by the artifact model and its results lead to changes in long-term memory researcher models.[6]

The analysis in this section argues for extending the notion of model-based cognition to comprise researcher and artifact models. Such (internal–external) representational coupling is a subcategory of the broader notion of "interlocking models." In general, problem-solving processes comprise experiments that generate, manipulate, and propagate interlocking researcher and artifact models within the wider cognitive-cultural system that is the lab. In the final section, we focus on the metaphor of "'interlocking" that we have used repeatedly throughout the chapter.

Interlocking Models

"Interlocking models" is a system-level interpretation we developed to capture dimensions of research practices that cut across many of our coding categories.[7] This multidimensional notion serves to articulate relations among the components of the laboratory cast as a cognitive-socio-cultural system. In particular, it provides a means of specifying in what ways the simulation devices serve as hubs. As with transportation systems, in which many service lines interlock at central stations, devices serve to interlock many dimensions of practice. In this chapter we have focused on three dimensions pertinent to articulating distributed model-based reasoning:

1. biology and engineering or interdisciplinary facets
2. researcher and artifact system components
3. configurations of models in experiments

As we have seen, experimentation in biomedical engineering requires that researchers create models by interlocking conceptual, material, and methodological facets of biology and engineering. Researcher conceptual models are hybrid representations, as are the artifact model-systems. In Lab A the primary interlockings are between cell biology and mechanical engineering; in Lab D at a minimum they include neuroscience, electrical engineering, and chemistry.

In the second and third sections of this chapter we provided numerous examples of interdisciplinary interlocking with respect to how researchers

---

[6] Christensen and Schunn's (2008) research on conceptual simulation and prototyping in engineering design showed that conceptual simulation decreased when prototypes were constructed, which provides an example of the latter case.

[7] See Kurz-Milcke, Nersessian, and Newstetter (2004); Nersessian (2005); Nersessian et al. (2003); and Nersessian and Patton (2009).

conceptualize and understand their model-systems. We reprise a few segments from transcript material from several Lab A researchers to illustrate the interlockings used in the construct and flow loop model-system. For a simulation to enable prediction about the in vivo phenomena, the flow loop and construct model-system needs to behave as though blood were flowing through an artery to a satisfactory degree of approximation. The flow loop, as a model, represents the shear stresses during blood flow in an artery to a *"first-order approximation of a blood vessel environment... as blood flows over the lumen"* and enables a *"way to impose a very well-defined shear stress across a very large population of cells."* Yet a flow loop simulation is *"something very abstract because there are many in vivo environments and many in vivo conditions within that environment. Things change constantly in our bodies over our lifetimes; including physiological flow rates."* The researchers make this approximation in design because *"as engineers, we try to eliminate as many extraneous variables as possible... so our conclusions can be drawn from the change of only one variable."* The original flow loop was designed to flow liquid over cells, but *"cell culture is not a physiological model"* and though much still is learned by just using cells, *"putting cells in plastic and exposing them to flow is not a very good simulation of what is actually happening in the body. Endothelial cells... have a natural neighbor called smooth muscle cells... these cell types communicate with one another."* The construct model *"behaves like a native artery because that's one step closer to being functional."* The constraints of the flow loop (designed originally for cells on slides and modified slightly for the thicker constructs) require that the tubular constructs be cut open and exposed to flow flat (unlike an in vivo artery). However, because the endothelial cells that line an artery in vivo are so small relative to the artery, the researchers believe that flowing the liquid (representing blood) over the flat in vitro construct (representing the artery) is a manipulation that will provide accurate enough information about the effect of forces on the endothelial cells: From the *"cell's perspective... the cell* [in vivo in an artery] *sees basically a flat surface."*

In just this brief exemplar of one model-system, the interdisciplinary interlocking models comprise the following elements, at least:

- biological and engineering models of in vivo phenomena in the wider community (as detailed in texts and journals)
- bioengineered in vitro artifact models
- researcher (conceptual) models of
  - in vivo and in vitro phenomena
  - devices qua in vitro models
  - devices qua engineered models

Finally, by adding to this account of interlocking models in distributed model-based reasoning processes an appreciation of the devices as "hubs" for interlocking the various dimensions of practice, we move farther along the path toward an integrative understanding of cognitive-socio-cultural dimensions of research practices. As hubs, devices are focal points around which lab activities revolve and through which they interlock. Devices and model-systems are what socio-cultural studies of science refer to as the "material culture" of the community, but are also what cognitive studies of science refer to as "cognitive artifacts" participating in the representational, reasoning, and problem-solving processes of a distributed system. Our data support their interpretation as cognitive-cultural artifacts. They are representations and thus play a role in model-based reasoning and problem-solving; they are central artifacts around which social practices form; they are sites of learning; they connect one generation of researchers to another; they perform as *cultural ratchets* (Tomasello, 1999) in an epistemic community (Knorr Cetina, 1999), enabling researchers to build on the results of the previous generations and thus move the problem solving forward.

SUMMARY

Given that the internal–external divide permeates the language of our culture, it is difficult to find language that adequately conveys the co-constitutive nature of culture and cognition that we believe to be mandated empirically and philosophically. We have used the language of partnering, coupling, and interlocking. In our choice of the first two terms, we have perhaps been influenced unconsciously by cultural conceptions of marital partnering as the "joining of two persons into one flesh." However far from the realities of marriage, the "two as one" notion does come close to expressing the kind of relationship of cognitive and cultural domains that enables them to be understood as a single system, each being intimately implicated in the other.

The need to solve problems not by experimenting on the in vivo phenomena, but through constructing in vitro models to simulate aspects of in vivo phenomena, leads to a complex form of reasoning, in which part of the reasoning process occurs in the mind and part in the real-world manipulation of the model. Extending the notion of mental models, we propose that the problem solving is best understood as achieved by a distributed cognitive system comprising conceptual models of researchers and real-world simulation models.

Finally, the notion of interlocking models captures the hub-like nature of the physical simulation models within the cognitive-cultural system. They are representations of current understanding and thus play a role in model-based reasoning; they are central to the social practices related to community membership; they provide the ties that bind generations of researchers; they are the cultural ratchets that enable one generation to build on work of previous generations.

4

The Feeling Person

In this chapter we consider the role of emotion in research practices by examining a class of expressions we have coded and analyzed as implicating *emotion, affect,* or *motivation* in the interview data. As noted, this book with its current focus was not planned at the time we began to analyze interview data, and we did not approach coding with the explicit goal of identifying expressions of this kind within the interviews conducted. However, while we were seeking to describe and code cognitive practices as evidenced in the interviews, we found striking examples of interview text that did not fit traditional cognitive categories, but seemed rather to have an affective or emotional tone. Other passages seemed expressive of desires, goals, and aspirations. Although not clearly cognitive, passages of these kinds seemed intimately related to and interwoven with accounts of problem formulation and solving, leading us eventually to code "affect/motivation" as a higher-order category and for this book to include it in the domain of general sense-making that is characteristic of laboratory activity. As Thagard remarked recently, "Affect is a natural subject for a dynamical theory that emphasizes the flow of thought and the complex interactions of emotion and cognition" (2008, p. 51). We also found that affective, emotional, and motivational expressions related intimately to identity formations, as is discussed in some detail in the next chapter.

Importantly, we assigned codes for affective and motivational expressions in both laboratories (A and D) and across levels of expertise, from PI to novice. Also noteworthy is that the expressions appeared not only in interviews intentionally focused on the researcher's experience in the laboratory but even in interviews aimed at obtaining descriptions of laboratory instruments or equipment. This finding was a principal source of inspiration for the idea that the laboratory has interesting psychological dimensions that cannot be exhaustively captured by existing social, cognitive, or material

categories. Further, emotion does not seem to receive detailed consideration even in analyses of human agency in science practice (e.g., Pickering,
1995). Although the attention we have given to the affective dimension of
laboratory practice has not been as extensive as our focus on cognitive practices per se, we see the affective dimension as illustrative of the way in which
a research laboratory provides a microcosmic demonstration of the complex integration of psychological functions, as argued in our introductory
chapter.

The task at hand is to illustrate what we mean when we characterize some
expressions as affective and motivational and how we see them integrating
with expressions more clearly indicative of cognitive practices. We do so with
examples from the text of different interviews, analyzing their function in the
context of each interview and of our observations of laboratory practice.
We begin with a brief discussion of the place of emotion in psychology;
then review some existing ideas about the relation of emotion to cognition,
rationality, and culture; and finally outline some conceptual challenges
related to any effort to talk about emotion with clarity.

EMOTION IN PSYCHOLOGY

It is common practice for general or introductory psychology texts to include
a chapter on emotion, sometimes in combination with motivation, but
usually in abstraction from the manifestation and function of emotions
in clinical symptoms. As with all of psychology's standard categories and
chapters, there is a story behind the convention of introducing students
to the study of emotion in just this way (see Danziger, 1997). However,
the practice of "covering" emotion(s) by reviewing standard theories and
research can greatly underplay the complications attending the lengthy
effort to understand what an emotion is, how an emotion develops and finds
expression, and how it relates to functions and experiences of other kinds.
For example, Parrott and Harré characterized emotions as "exemplary of the
ways in which contemporary psychology faces an irreducible complexity.
Emotions are at once bodily responses and expressions of judgments, at
once somatic and cognitive. They seem to have deep evolutionary roots, yet
they are, among human phenomena, notably culturally variable in many of
their aspects" (1996, p. 1). Ronald de Sousa declared emotions to be "like
Descartes' pineal gland: the function where mind and body most closely
and mysteriously interact" (1997, p. xvi). They "seem to overstep a threshold
of messiness beyond which even the most masochistic of theoreticians tend
to lose heart" (de Sousa, 1995, p. 270).

The complexity involved is underscored not only by a deep history of philosophical interest in emotion but also by advances in the neuroscience of emotion,[1] as well as a proliferation of accounts emphasizing emotions' fundamentally social, cultural, and/or political production.[2] Of course, part of the difficulty lies in the great variety of terms used to refer to this class of experiences and functions. Feelings, desires, emotions, moods, passions, pathos, and being "affected" by objects can be conceptually and historically distinguished; the term "emotion" is used in reference to both fleeting states and dispositions. Moreover the conceptual relation between emotions and other psychological categories has varied widely over the decades and by academic perspective. For example, a recent and typical general psychology text defined motivation as "the set of factors that activate direct, and maintain behavior, usually toward some goal.... Emotion refers to a subjective feeling that includes arousal (heart pounding), cognition (thoughts, values, and expectations), and expressive behaviors (smiles, frowns, and running). In other words, motivation energizes and directs behavior. Emotion is the 'feeling' response" (Huffman, 2010, p. 406). By comparison, William James earlier reproached empirical psychology for focusing on the "perceptive and volitional" at the expense of "the *aesthetic* sphere of the mind, its longings, its pleasures and pains, and its emotions" (James, 1884, p. 188, emphasis original). Emotion also has been conceptually distinguished from desire (Wollheim, 1999).

The rather convoluted route by which psychologists have related the categories of emotion and motivation has been documented in many textbooks on motivation. For several decades motivation was distinguished from emotion and theorized in terms of drive (e.g., Franken, 2006). Motivation was also sometimes linked to personality, notably by Maslow who related it to the dispositional tendencies and potential capacities of any person (1954/1987). Personality, in turn, linked back to emotion, through long-term and characteristic patterns of emotional responding. But so much remains in a muddle.

EMOTION AND RATIONALITY

Among the perpetual questions pertinent to the nature of emotion is the relation of emotional states to the forms of activity we classify as rational.

---

[1] See, e.g., Davidson and Sutton (1995); LeDoux (1998); and Damasio (1994, 1999).
[2] See, e.g., Gross, (2006); Harré (1986); Lutz and White (1986); and Parkinson, Fisher, and Manstead (2005).

At stake is one version of the mind–body problem, which is ancient, many-sided, and hopelessly complicated by the imprecision of language.

The ideal of reason as dispassionate and uninvolved, by which it is empowered to serve as a platform for jurisprudence, has long stoked a popular conception of the "heart and head" as in oppositional relation. "The standard view of the relation between rationality and emotion is, of course, that emotions interfere with rationality. They are, as it were, sand in the machinery of action" (Elster, 1996, p. 1394). According to de Sousa, the distortion of judgment by emotion is "what all good novels are about" (1997, p. 197). Philosophical support for the view that emotions impede rational powers is typically assumed to originate with Plato's *Phaedrus* (Solomon, 1993; Thagard, 2008), through the model by which persons are internally conflicted by the demands of opposing agencies: "a committee of subpersonal persons" (de Sousa, 1997, p. 27). Although Aristotle was to portray persons as equipped with an array of capacities or functions rather than parts, thereby promoting a more nuanced view of the emotion–reason relation, including the place of emotion in moral reasoning (e.g., see Konstan, 2006; Robinson, 1989), an ideal of unsullied rationality reached its fullest expression in stoicism.[3] Descartes, Spinoza, Hume, Hobbes, and Kant took up the emotion–rationality relation in varying ways and with different degrees of emphasis. More recently, artificial intelligence has promoted a view of reason as the manipulation of abstract symbols that are "meaningless in themselves" until meaningfully attached through reference to actual or possible things, thereby giving them endless combinatory possibility (e.g., Fodor, 1980). The controversial implication is the existence of a realm of pure symbolic cognition/reason (categories), prompting critiques of "reason as the *disembodied* manipulation of abstract symbols" (Lakoff, 1987, p. 8, emphasis added).

Reviews of empirical literature investigating the relation between reason and emotion support the ancient assumption that there is a disruption to reasoning occurring under conditions of negative emotional states of strong intensity. However, more mild negativity actually seems to enhance certain kinds of reasoning tasks, such as systematic thinking, reappraisal, and attunement to data: a condition dubbed "depressive realism" (Alloy & Abrahamson, 1988). Effects on reasoning of both high and lower intensity positive emotional states are less well understood (see Ben Ze'ev, 2000;

---

[3] Even here, however, the nature of the relation between emotion and rationality, including emotion and wisdom, has been oversimplified. For an excellent recent discussion on this point see Graver (2007).

Pham, 2007). Another strongly supported trend is the implication of emotion in moral reasoning (i.e., that reference to one's own emotional experience provides the grounding for judgments of conditions or actions as worthy, unjust, or taboo; Haidt, 2001, 2007; Prinz, 2007). However, reflecting the conceptual confusion surrounding the use of the terms involved and the distribution across a range of disciplines – cognitive psychology, social psychology, economics, and neuroscience – a recent review of the status of empirical findings on the emotions and rationality (or "affect and reason") proclaimed the literature on this connection to be "very fragmented and seemingly inconsistent" (Pham, 2007, p. 155).

A comprehensive statement of the involvement of emotions in rational functioning is de Sousa's claim that emotions do not merely interact with or *influence* reasoning but "*underlie* our rational processes," moral and otherwise (1997, p. 201, emphasis added). Purely logical operations are frequently incapable of determining what is most salient to the problem at hand; thus the very fundamental business of determining where to direct attention and when to engage in further inquiry has an essentially emotional core. Because we "know" too little to base decisions about pursuing inquiry merely on our knowledge of the way the world works, emotional responding serves as a directional short-circuit for inference and decision. De Sousa termed this function of emotion "intuitive" because the basis for the shifts of attention is frequently not subject to explanation or even articulation. On the other hand, noted de Sousa, artificial intelligence has alerted us to "a class of problems generated by the fact that we know too *much*" (p. 192, emphasis added). That is, in planning any action (itself a form of acting), an almost infinite number of facts are available as a resource; of these we must make use of only a small subset. The basis for these choices, as well as for the choice between two desirable strategies in a scenario driven by the basic "rational" principle of "maximizing expected gain" (Bayesian decision theory), was attributed by de Sousa to emotional or dispositional factors. Emotions serve a heuristic function, reducing the need for deliberation among all possible alternatives. Therefore, de Sousa and others (e.g., Nussbaum, 2001) have emphasized that emotions have a kind of inherent rationality or intelligence.[4]

---

[4] Note that although there is some overlap, the idea that "emotions are intelligent" should be distinguished from the concept of "emotional intelligence" (EI) defined by "the competence to identify and express emotions, assimilate emotions in thought, and regulate both positive and negative emotions in the self and others" (Matthews, Zeidner, & Roberts, 2002, p. 3).

EMOTION AND COGNITION

The conceptual relation between reason/rationality and *cognition* has never been settled firmly, and thus it is more difficult to characterize enduring or popular conceptions of the relation between emotion and cognition. In any case, in recent decades there have been analyses of both the cognitive basis of emotion and the emotional basis of cognition, leading to new questions concerning the clarity with which these processes can be distinguished meaningfully. Interest in emotion and its role in cognitive processes has prompted an "affective revolution" comparable in some respects to the "cognitive revolution" of the 20th century (Haidt, 2007, p. 998).

Cognitive Emotions

After a period of neglect in the early 20th century, a philosophical reassessment of emotion occurred that placed new emphasis on the propositional content of emotional states – their status as having something to say *about* the objects or situations toward which they are directed – leading to a conception of emotion as strongly cognitive (e.g., Bedford, 1962; Solomon, 1976; see also Deigh, 1994). Theories emphasizing the basis of emotion in the (cognitive) *appraisal* of events and situations began to be developed in earnest in mid-century and have been influential for decades (Arnold, 1960; Lazarus, Averill, & Opton, 1970; Roseman, 1991). A variation is to regard emotions as judgments (Neu, 2000) or even choice (Solomon, 2003). The conception of clinical symptoms and behavioral problems as rooted in acquired habits of maladaptive thinking soon followed, thus prompting the assumption that symptoms could be cured or reversed (unlearned) by altering cognitive patterns (e.g., Beck, 1970; Ellis, 1977).

Emotional and Embodied Cognition

More recently, the profound impact of emotion on cognitive tasks and processes has been argued compellingly, prompting the appearance of the rather intriguing term "emotional cognition" (Moore & Oaksford, 2002; Thagard, 2008). Thagard has aimed to increase awareness of the mechanisms through which emotions and motivations affect decision making, both positively and negatively, toward the end that we might harness emotional power and achieve greater effectiveness in domains of practice from law to religion to science. An example with potential application is the evidence from studies of brain lesion and injury that emotional or somatic marking contributes to our ability to learn from our mistakes (Bechara, 2004; Damasio, 1994, 1996).

Because even cognitive theories of emotion acknowledge a somatic or bodily feeling component to emotional experience, the literature on emotional cognition is more broadly situated within the extensive recent literature on embodied mind and embodied cognition (Johnson, 2007; Lakoff & Johnson, 1999; Varela, Thompson, & Rosch, 1991). Although emotion is not reducible to bodily experience, "[i]n a rough and ready way, emotions are assumed to fall entirely on the 'body' side of the 'mind-body distinction' for anyone who would allow that much talk in mentalistic terms" (Greenspan, 2003, p. 114). Viewing emotional cognition as part of embodied cognition perspectives implicates emotion even more deeply and fully. If cognition is admitted to be fundamentally embodied, of course, the door is opened to a floodgate of emotional influence on all levels of cognitive functioning. As one example, an important question concerns the extent and nature of emotional involvement in the metaphorical foundations of our conceptual systems, even mathematical systems (Lakoff & Nuñez, 2000). The close association of metaphor and emotion has been well documented in cognitive linguistics (Johnson, 2007; Kövecses, 2000; Lakoff & Johnson, 1980).

Emotion's involvement in cognition emerged in relation to discussions of high-level skill acquisition. Dreyfus and Dreyfus (1979) analyzed models of instruction designed for training pilots to respond to emergencies. They concluded that training through the use of richly detailed case examples is most effective because it incites imaginative and emotional involvement: "This richly realistic detail presumably encourages the trainee to become so involved that he creates *emotion-laden images* and thus builds up his own paradigm of each type of situation" (pp. 30–31, emphasis added). In their later (1980) presentation of a five-stage model of the cognitive basis of skill acquisition (novice, competence, proficiency, expertise, and mastery), Dreyfus and Dreyfus incorporated "moments of intense absorption" into the description of the highest level of cognitive function, made possible by a relaxing of conscious attention to the task at hand (p. 14). In this paper and in later work by Hubert Dreyfus (1997), the intense absorption is depicted as a channeling of mental energy accompanying an "intuitive" state of mind that operates interactively with an analytic framework, rather than an explicitly emotional state. Yet it is difficult to imagine that "intense absorption" is outside the affective or emotional domain.

EMOTION AND SCIENCE

In contrast to the mounting evidence pointing to intimate connections between emotion and cognition, there is comparatively little analysis of the role of emotion in scientific practice, though Manuel's psychoanalytic study of Isaac Newton (1968) and John Bowlby's attachment theory account of Darwin (Bowlby, 1991) might be cited as important examples. Other historically important exceptions are noted in the final chapter of this book, in which we situate our analysis in a broader and deeper tradition. One worth noting here is Polanyi, who devoted a chapter of *Personal Knowledge* to the topic of "Intellectual Passions." It made bold claims about the "manifest emotional force" of scientific process that is most obvious in the discovery phase, including the delight attending acts of ingenuity and the joy of contemplating natural mysteries, the intellectual beauty of mathematics, and the appreciation of elegance (Polanyi, 1958/1973, p. 133). Likewise, a wide range of theorists, historic and contemporary, have acknowledged the aesthetic experience of scientific theories within the general parameters of scientific values (Kuhn, 1977; McAllister, 1996). Similarly, Mitroff (1974) and Mahoney (2004) included emotion in their analyses of the contribution of subjectivity to the scientific process.

However, emotion principally enters the discourse of contemporary science studies through biographical or autobiographical studies of particular scientists, studies of scientific creativity, and analysis of the personality characteristics of scientists as a group, classified into cognitive, social, and motivational traits (Feist, 2006a, chapter 5; Simonton, 2004). In contemporary cognitive science, Thagard's chapters on scientific cognition in his recent *Hot Thought* (2008) have provided one of the few extensions of recent emotional cognition literature to the domain of science. To analyze the role of emotion in all phases of inquiry, Thagard used historical case data relating to Watson and Crick's discovery of the structure of DNA and Dr. Patrick Lee's discovery that the reovirus has the potential to kill cancer cells. His analysis, which was more detailed in relation to Watson, offered "a rich set of examples of possible emotional concomitants of scientific thinking" (2008, p. 173). Thagard began by identifying "emotion words" in the 143 pages of *The Double Helix*, Watson's "personal account" of his famous discovery with Crick (Watson, 1969). Thagard found a total of 233 emotion words: 125 attributed to Watson's own experience, 35 to that of Crick, 13 in reference to Watson and Crick in combination, and 60 in relation to other researchers.[5] He noted the problem of determining the accuracy of the emotions that Watson attributed to Crick or the other researchers. Thagard's next step was to code the emotion words as having either positive or negative valence, and he found more than half to have positive valence. He then coded the emotions to align with the categories of "basic" emotions assumed to be universal: happiness, sadness, anger, fear, disgust, and surprise. He identified additional words that had emotional tone but did not fit these basic categories: interest, hope, and beauty. Happiness appeared most often, followed by "interest," and "wonder and enthusiasm" both at the joy of discovery and at the elegance of the model defended. However, negative emotions – sadness or anger at the failure of a research effort and anxiety over potential failures or competition with other researchers – were also common and seemed to be motivating in function. Finally, Thagard evaluated the place of each emotion word in the different phases of inquiry. He began with Reichenbach's basic distinction between the contexts of discovery and justification and then drew a further division between investigation and discovery to account for the extensive work that frequently precedes discovery. Thagard found that most of Watson's emotion words took place in the context of investigation, but identified emotion in all three phases, including justification. In making sense of his coding, Thagard questioned the appropriateness of the models of rational decision making derived from economics, particularly as framed in terms of maximizing epistemic gains (e.g., truth), to account for the choice of research topics: "It is rarely possible for scientists to predict with any accuracy what choice of research areas, topics, and questions will pay off with respect to understanding, truth, or personal success" (p. 175). Moreover, "once interest and curiosity direct scientists to pursue answers to particular questions, other emotions such as happiness and hope can help motivate them to perform the often laborious investigations that are required to produce results" (pp. 175–176).

According to Thagard it is much more controversial to claim a strong basis for emotion in the context of the justification phase, given Reichenbach's view that "subjective factors" were operational in the discovery context only. Yet Thagard cited references to the aesthetic appeal of theories as frequent markers of their acceptability both in relation to Watson's work and as analyzed by others (e.g., McAllister, 1996). Even more broadly, Thagard argued that judgments of the coherence of theories hinge on emotional experience: "attachment to a positive or negative emotional assessment to a proposition, object, concept, or other representation" (p. 182). Extending

---

[5] 235 in Thagard (2008), but the numbers he provides add to 233.

this analysis further, he viewed conceptual change as invoking emotional change.

EMOTION AND MOTIVATION IN BIOMEDICAL ENGINEERING LABORATORIES: OUR ANALYSIS

Our analysis of emotion and motivation in science practice bears some similarities to and some important differences from that undertaken by Thagard (2008). Similarities include the identification of "emotion text," which in our case are passages of interview text that we interpret as expressions of emotional experience. Like Thagard, we examined this text's valence, positive or negative, and the target of the attribution; that is, to whom (or what) the emotion is attributed. Unlike Thagard, however, we focused on the presence of emotion expressions in the context of *interviews* with scientists conducted by our research team. We were thus able to analyze the function of emotion expressions in a more immediate account of science practice rather than an edited, published account. Second, we included attributions of emotions to *objects and artifacts* important to laboratory practice. Third, we did not take assumed universal categories of emotion as a guide. We coded in an open-ended fashion, attempting to describe the emotional or motivational theme of the passages as seemed most fitting. Fourth, we examined not only emotion words but also longer passages of text suggestive of emotional or motivational states, as well as figurative language that appeared to have an emotional tone or function when interpreted in context. Finally, because our interviews had different foci, we did not analyze the expressions in terms of the stage of problem solving in which they occurred (as did Thagard); rather, we considered the overall function of the expression in the particular problem-solving context as recorded in the interview.

As noted, our analysis of emotion is still in an early stage, and we have conducted fine-grained coding on only a small subset of our large collection of interviews. In this chapter, we provide examples of different forms of expressions of emotion or motivation and consider how they relate to other forms of sense-making indicated in the interview event. Our emphasis is not on counting the number of emotion words or expressions but on characterizing the variety of functions such expressions play in the context of the interviews and illustrating these functions with examples. Where possible we attempt to provide examples from a variety of researchers in both laboratories. We organize our examples into three categories of emotion-related expression: overt expressions, figurative or metaphorical expressions, and anthropomorphizing expressions. Our examples illustrate not only the close relation of emotion to problem solving and dedication to the work of the laboratory but also implicate social and cultural dimensions of emotion expression. In particular, we found that an interactional or transactional theory of emotion became most salient to cognitive practice in relation to anthropomorphisms. Therefore we conclude this chapter with a discussion of the implications of this category of expression.

OVERT EXPRESSIONS OF EMOTION OR MOTIVATION

Many of the interviews conducted for the purposes of understanding cognitive practices in the BME laboratories focused on the problem-solving and goal-directed actions of the researchers interviewed. It is not surprising, therefore, that statements expressive of motivational dimensions of practice made frequent appearances in interview transcripts. Although conceptual distinctions between motivation and emotion are fuzzy, their close connection to problem solving is obvious in the role emotions play in both facilitating and undermining progress toward acknowledged goals. We found various passages expressing frustration in relation to obstacles in problem solving and others expressing positive emotions in relation to a research program or career path going well. We offer brief comments on the theoretical significance of the passages we cite as analyzed within the context in which they occurred. Of particular interest is the way in which both excitement and frustration are associated with the open-ended, cutting-edge nature of the research undertaken in these laboratories, which we have elsewhere identified as *agentive learning environments* and as *innovation communities*. We therefore include in our discussion of overt expressions of emotion a look at how they appear in relation to those features of laboratory research.

Excitement

We begin with a general look at the place of excitement in our transcripts. Here in an interview focused on research progress, a student working in Lab D says, "*Right now **it's not too exciting**. (I'm) taking courses and I haven't been able to do as much as I like as far as research or anything like that*" [2002–03–25-D-i-D2]. Note that in this example, "*not too exciting*" is equated with not doing much research. An interview conducted 9 months later finds this same student still expressing motivation in terms of the chance to participate in research:

> And I've developed a lot of theory but I haven't had the chance to do anything yet. . . . **I am really anxious to do experiments**. . . . I've never really done

experiments before, so I'm really curious to see how good I'm at it. I'm just kind of delaying picking up more and more theory [2002–12–05-D-i-D2].

For D2, excitement and motivation are connected to his own "doing," his own participation in experimental science. The second passage implies a contrast between the active research participation D2 imagines as exciting and the accumulation of theory abstracted from hands-on practice.

The focus of excitement is quite different for D6, who in giving the interviewer a laboratory tour expresses excitement both about a new microscope and the use of "real robots":

*So the latest addition to the lab **I'm very excited** about, the new microscope. . . . One of the **exciting** things that we've gotten involved in is using real robots instead of simulated creatures* [2002-12–02-D-D6 (Tour)].

After identifying the use of real robots as exciting, D6 links his excitement to the *theoretical* import of the robots for their role in promoting an understanding of creativity:

And, so this year we had this thing [robotic arm] *being controlled by neurons in our lab here right here, so we had a link, a real-time feedback link going halfway across the world to Z35 . . . and the neurons would continue drawing some drawing. . . . I think this is important for a number of reasons. One of them, is that **it gets people to think about what is the minimum necessary for a creative process to happen**. You know, what is the process of creativity, and the whole process of artistic creation. Another thing is the integration of human-made artifacts and biological tissues* [2002-12–02–D-D6 (Tour)].

The use of robots is exciting precisely because of its theoretical significance, which includes reflection on the minimum requirements of creativity.

D6 also expresses excitement in the context of a narrative about his entry into the field of neuroscience:

*So, so uh, when I went to college, I didn't know what kind of scientist I wanted to be. My roommate said, "You ought to come to this class that I'm taking. It's about cognition." . . . I accompanied him to class and sure enough **everything he said was absolutely fascinating to me**, and I went in to go, to go chat with him in his office hour. . . . **In that half an hour it was fantastic, what he said was just wonderful, it sucked me in**. . . . So, that was **extremely exciting to me** and he was at Z12 at that time, and I made a road trip up from Z10 to Z12 to go and hear him give a talk and got even more inspired. And at that point I had decided, ok, neuroscience is what I want to do, and I wanna be a neuroscientist.*

This passage clearly relates the excitement D6 remembers experiencing in first being exposed to a set of ideas about cognition and then deciding to pursue neuroscience as a career.

Participants sometimes attributed expressions of excitement to other researchers. For example, in a progress interview, A22 cannot remember the details of A10's experiment, but she remembers that A10 was very excited about it:

> I: So have you learned any new procedures this week?
>
> A22: *No, not really. Um, I've gotten to see the results of some new stuff that, that um, that A10 did. I don't – can't tell you much about it because I didn't have a chance to learn.*
>
> I: What did A10 do?
>
> A22: *Um, he – see, this is what I'm going to have to try to remember – um, he made tissue – some tissue engineered. . . . I don't even remember what it is. It has something to do with the heart valve that he was working on. . . . **But he was very excited** [laughs] [2002–11–13-A-i-A22].*

Frustration

We additionally found many references to frustration, both overt and implied, attributed by researchers to themselves and to others.

For example, A31 repeatedly expresses unpleasant emotion in relation to the task of graphing the data in Lab A: "*I'm **still struggling** with the issue of graphing the data.*" This struggle affects his work by limiting the amount of time he can bring himself to devote to the project: "*I get **real frustrated**, so I only work in increments.*" He elaborates with the interviewer's prompting:

> *You have to set your array size and then initialize it. So I figured out how to do that, but then I can't figure out how if, if I have a 1 by 10,000 element array, it's empty, um, I can't figure out how to append to the end. . . . So I got **pissed off** about that, and I started working on the linear encoder* [2005–02–02-A-i-A31].

In other words, A31's frustration has resulted in a change of task; he switched to working on the linear encoder. Several weeks later he tells another interviewer that he has

> *taken enough time off from this thing that I'm not pissed off at it anymore. . . . There was a period there where I was working on it everyday and I was just like . . . really . . . you know . . . I'm going to lose it if I can't get this to work* [2005–02–22-A-i-A31].

Connections between Emotion and Motivation

Emotions also enter the narratives researchers provided concerning their entry into biomedical engineering and their career paths. For example, negatively charged expressions color the account D6 provides of his graduate school experience: He describes his relation with his own academic advisor as "*hell*," explaining that when he was merely learning techniques he and the advisor got along well, but that "*at some point when I, when I learned enough to start coming up with my own ideas of how we should do things, we really butted heads.... And when I have ideas I expect that he should at least consider them but often he did not even consider them and I was **really grinding my teeth**"* [2002–12–11-D-i-D6].

Frustration is also attributed to other researchers, as in this passage by A22: "*A54 – I don't even know his last name... I was talking with him yesterday, and I know **he's frustrated** because he can't understand all the jargon*" [2002–10–10-A-i-A22].

Anger is also attributed to a computer, in the context of the researcher not being able to demonstrate a point to the interviewer by pointing to a display on its screen: "*It's alright. It's buggy sometimes. It **gets angry** when it has too much stuff*" [2005–02–02-A-i-A31].

Most frequently, however, expressions of frustration occurred in conjunction with accounts of failed experiments or work not progressing as planned. Given the cutting-edge nature of their research, failure is a daily fact of life in these labs. In this example, D4 frames her current paper-writing effort in the context of her overarching motivation to perform a plasticity experiment:

> We're writing an experimental paper about how to suppress bursts.... Straightforward.... But my whole motivation for ever getting into suppressing bursts is plasticity experiment, ok? I spent March to August last year trying to get this work – plasticity experiment – working. OF COURSE it didn't [2003–04–09-D-i-D4].

Although the researcher does not express frustration directly in this passage it is suggested by what seems to be an ironic "*OF COURSE it didn't*" (work) at the end of the passage. Following up, the interviewer merely repeats, "It didn't?" to which D4 gives a lengthy and seemingly emotionally charged reply:

> It did NOT. Very good, so I said why is it not working? And I thought, ahhh! The bursts! Ok, then I spend the rest of the year trying to do, till this March, trying to do burst suppression. Succeeded, so called. Succeeded... and now,

this March, I'm back again, to plasticity experiment [starts laughing]. *So it's pretty interesting, I've spent an entire year doing something else, and I've come, it's around the **same damn time** that I'm doing **the same damn plasticity experiment**. But, now I feel that this should work. **If it doesn't, then it's very bad. It's very bad.** So, every time I go and start this plasticity experiment, I say, "**my God, if this doesn't work**"* [pumping fist], *my **whole hypothesis, one and a half years of work can go into the dustbin** and it's just, fine, so, OK; I'll have to try a totally new line of thought* [2003–04–09-D-i-D4].

This passage vividly illustrates the unpleasant aspects of science practice that counteract the thrill of innovation or the vigor accompanying theoretical challenge. Despite identifying plasticity experimentation as her "*whole motivation*," D4 later marks it and the time she devotes to it with the expression "*same damn.*" The pressure she appears to feel to obtain her desired result is indicated in her statement that her hypothesis and 18 months of effort will be wasted ("*can go into the dustbin*"). Of interest is that the interviewer noted D4's pumping fist, even though the interview was conducted during the early stage of our investigation when we had given no indication that we would be looking at the function of affect or emotional expression in the laboratories.

Despite her frustration, however, D4's account also underscores what has been increasingly emphasized in recent analyses of the role of failure in innovation, including scientific discovery (e.g., Fetzer, 2003; Petroski, 2008). If her hypothesis goes into the dustbin, D4 will embark on a new direction of thought and experimentation. The unpleasant experience of frustration evoked by failing experiments may be a necessary impetus to innovation or heightened efficiency. This idea is also expressed by a Lab A researcher in relation to his efforts to obtain desired results:

*So I tried to do all this stuff like build array. You know and you use, set your whatever element array.... But for some reason it just wouldn't work.... **It seems like it should work. And that's what really frustrated me**, and so I started messing with this auto-indexing thing* [2005–02–02-A-i-A31].

Note that A31 expresses frustration in part because the results of his manipulations do not accord with what he thinks is logical, what makes sense to him. He begins to discuss a new course of action, "*messing with this auto-indexing thing*," immediately after acknowledging his frustration. In other words, there appears to be a relation between the frustration evoked by the failure to obtain the expected result and the new action taken to remedy the problems.

This view of frustration as an unpleasant but important spur to new action accords with an interesting finding emerging from our coding in relation to what we have framed as the *agentive learning environment* of the laboratories – that is, environments in which graduate and undergraduate students are empowered to be agents of their own learning – as well as to their status as innovation communities. These features of the laboratory communities are associated with both positive and negatively charged emotions. In the following, A7 and D7 acknowledge the pleasant aspects of working independently, but the tone of the passage from an interview with D4 is decidedly more negative.

In an interview ostensibly devoted to a description of the physical simulation models they design and construct, locally called "devices," A7 calls working in Lab A

> **a fun adventure . . . you fumble along,** *and hopefully you're better off you know for having taken time to really think through what you're doing . . . so I mean* **you totally have ownership of this . . . you know at the end of the day that would feel good** [2004–07–01-A-i-A7].

Similarly, D7 acknowledges the potential for some of Lab D's "big ideas" to fail, yet calls the research "fun" and speaks of his love of being part of the lab:

> D7: *And these are very, very big ideas. . . . A few of them might, um, might fail. You know.*
> I: *But that's research as you said.*
> D7: *So like,* **I take it as fun, I love, I love to do, I love to think like this.** *And that's why I'm here. Because, here I know that I'm interacting with* [the PI] *for a long time, and I know that I can get, get good opportunity here, like, this will be mutual. If I, I'm able to join in this lab, they will also benefit, I will also get satisfaction* [2002–10–24-D-i-D7].

In contrast, D4 associates the knowledge-at-the-frontier status of research in Lab D with "trouble," and the affective tone of this description of her experience is at least mildly negative:

> *But there are a lot of questions and nobody knows how to answer them. Nor does [the PI]. . . . Yeah, I don't know what's happening. How much do I know what's happening? Very little. There's, there's a lot that I just don't know. . . . So, the deal is there are too many questions and very few answers, and* **that's where the trouble is** [2003–04–09-D-i-D4].

In these examples, the close relation between the emotional expression and the account of theoretical innovation and either progress or disruption in research progress leads us to include emotion and motivation in the broad category of sense-making we see as primary activities of the acting person. Importantly, however, emotions also relate intimately to what we are describing here as identity formation, which we discuss in greater detail in the next chapter.

FIGURATIVE AND METAPHORICAL EXPRESSIONS

However, as interesting as the clear expressions of emotion and motivation in the interview transcripts prove to be, there is a need to look beyond them to fully appreciate the role of emotion in laboratory practices directed toward the production and dissemination of knowledge. This role is best understood by considering theories that emphasize the social nature and function of emotion. For example, Kövecses (2000) argued against the strong biological position on emotions that principally implicates brain states rather than linguistic distinctions in the evolution and experience of emotion (LeDoux, 1996). He considered conscious feelings to be vitally involved in emotion and noted that these feelings "are often expressed in or, indeed, are shaped by language" (Kövecses, 2000, p. ix). What is required to understand this connection deeply is an abandonment of what he considered "simplistic views of emotion language" (p. xii), in which "emotion language will not be seen as a collection of literal words that categorize and refer to a preexisting emotional reality, but as a language that can be figurative and that can define and even create emotional experiences for us" (p. xii).

In our dataset, then, as a second class of emotion expressions we consider the function in context of what we call *affectively toned metaphorical expressions*. These are sprinkled throughout the interviews and are more subtle forms of emotional expression than are the overt examples already discussed.

Little Beasts and Violent Methods

In this interview about laboratory equipment, D1 refers to a vibrator as a "little beast" for no obvious reason related to the interview context other than that the vibration is of high frequency. Calling a piece of equipment a "little beast," however, suggests a feeling that it is wild and uncontrollable – and possibly a sense of danger.

> 1: So I noticed the hood was cleaned out, from stuff. Just wondering what was going on there.

D1: *What they're doing with this right is platinizing MEAs. . . . And this* **little beast** *is a, um, it's a high-frequency vibrator is essentially what it is, and it produces a high-frequency vibration* [2002–12–05-D-i-D1].

In the same interview, D1 offers this appraisal of Lab D's methodology:

I: So how do the methods change with the technology?

D1: *The technology seems to be racing forward dramatically, and, um, the methods are slowly beginning to catch up. But, you know, from what I've seem, cause I come from the cognitive learning and memory area, and we have a long history of experimental method . . . with lots of different types of controls you can do for different types of phenomena. . . . From what I've seen in this field so far, the method seems to be very simple; it's a typical A-B-A type of preparation. . . .*

I: You think the methods in this field are too deeply rooted and maybe need to grow?

D1: *They will over time, because they'll make the same sorts of mistakes that we did using those types of methods. I mean, it's bound to happen and make the mistakes that will take you down the wrong path, and you'll get better and doing what you do. My advantage is that I know about all this stuff already. So I know how* **I can be killed** *by this type of procedure* [2002–12–05-D-i-D1].

It is not in any way obvious how A-B-A (sic) research design is in and of itself a lethal procedure; thus we can assume that the metaphoric expression signals some frustration or disappointment D1 suffered while conducting a prior experimental procedure. "*Being killed*" expresses a sense of victimization. In contrast, A31's language in the following is suggestive of more violent impulses toward a measurement tool:

I: The measurement you are getting off there . . . I guess the red line . . . it is independent of what you are trying to produce? Independent of the voltage you are sending out to control it?

A31: *Yes I think so. When it was set up incorrectly and I command it to move 5000 counts that way, right? A lot of times it would do something like this and shoot the wrong direction and oscillate out of control . . . it was nuts . . . it sounded like a jack hammer . . .* [both laugh] *and* **I had to press kill, kill, kill!** *It needs a big red button to control the damage* [2005–02–22-A-i-A31].

What we are claiming is that such figurative or metaphorical language provides a subtle indication of affect that is easy to overlook. Yet the affective coloring is significant in that it is integrated with the cognitive practices that are also in evidence (e.g., describing equipment or methodology). That these expressions have affective coloring is underscored by comparing them to other metaphorical expressions appearing in the transcripts that call on metaphor principally for the purposes of explication and clearly lack such coloring:

> I: And so what other things might you look at on – what else to see the results of the bioreactor?
>
> A16: *And so the elastic, **the elastic part is similar to – you think about a rubber band**, a rubber band can recover if you just stretch it back and forth, and back and forth, and it doesn't seem to change in any way* [2002–07–15-A-i-A16].

> A10: *So like, a blood vessel, the matrix is collagen. Cells sit in there. So you have these – **it's like it's putting rocks in jello**. So the jello's the matrix and the rock's the cell* [2002–07–05-A-i-A10].

> A10: ***Basically turbulent flow is like in the back of a canoe paddle with the eddies*** [2002–07–09-A-i-A10].

In these examples, the use of metaphor and analogy appears to be deliberate and for the purpose of enhancing the interviewer's understanding through visualization. Thus our claim is not that all metaphorical or analogical expressions are affective in tone, but that in some cases metaphor seems to signal affectively toned expressions. More broadly, metaphorical expressions might be a source for identifying subtle contributions of the affective dimension to sense-making practices in science: "There is very little about the emotions that is not metaphorically conceived" (Kövecses, 2000, p. 85).

"Cool" Science

Also easily overlooked as an affective marker is the multivalent term "cool," which makes frequent appearances in our interview transcripts. Its meaning, though figurative, is largely indeterminate, and the word is used in a wide range of circumstances in ordinary discourse. Certainly it is not one of the basic categories of emotion and might be disregarded in an analysis focusing on these categories. Indeed, it is tempting to dismiss the frequency with which various aspects of laboratory practice are called "cool" by our researchers as reflective of the ubiquity and banality of "cool" in casual American discourse. However, read in the context of the discussion recorded in the interview transcripts, "cool" marks emotionally laden and motivationally relevant expressions of several varieties. Note that in these examples the use of "cool" flags particular interest or involvement in relation to what are clearly cognitive practices.

The connotation of "cool" in the interview context is invariably positive, associated with something desirable. It is used in relation to several different kinds of situations/events: career prospects, a research problem, an engineered product (robot), an experimental outcome ("*so that would be really cool if that happened*"; [2003–04–09-D-i-D4]).

D4 calls Lab D's robot "really, really cool" because of its affordances for problem solving and innovation:

> *I don't know whether you've heard of [the PI's] animat thing which is basically this simulated animal.... The point is to have an embodiment for these cultures which is basically a robot.... If we get some recording patterns of this dish that we will be able to do something with... So... **that robot is really, really cool**. It has sonar systems* [2003–3–30-D-i-D4].

The excitement D6 expresses in relation to the acquisition of a new microscope is framed in terms of the *aesthetic* appeal of the possibilities it affords. Here "cool" is also paired with "exciting":

> *After a year or so this thing was ready, and it just worked the very first time we tried it out to put a post-it note under the microscope it, it made this* [emphasized] **gorgeous picture of paper fibers... Wow, this is so cool... very exciting** [2002–12–11-D-i-D6].

"Cool" also implicates *value* in science practice, as its presence in an interview passage serves to demarcate worthy from unworthy things. For example, D4 contrasts "cool" with "not clever" in the following passage: "*So, my cluster may actually accounts for the slant so, it's basically an expectation maximization kind of algorithm.* **It isn't any cool thing** except **the clever thing** *I did is with this, put some parameter which accounts for the slant for the shift in latencies*" [2003–02–27-D-i-D4]. "*Really, really cool*" marks something with conceptual clout, as indicated when D6 reflects on the potential of the robot's "creations" for helping scientists think through the minimal requirements for creativity.

By contrast, in the following example, "cool" is closely identified with the potential *social* relevance of scientific practice:

> I: What do you think you'll do when you finish (graduate school)?
> A7: *Industry is what I've always kind of had on the top of the list.... I think some of the* **cool** *aspects I idealize about industry, whether it's true or not, is being hopefully close to the clinic (sic) ... in other words, hopefully working on things that are very close to moving into um* **aiding human health**. *... I mean you think um a lot of basic science is* **really, really cool** *in terms of what we're learning and* **how we can apply it** [2004–07–01-A-i-A7].

The frequent appearance of "cool" in our interview transcripts is reflective of the interplay of cultural influence on emotional expression. The term, firmly rooted in the 20th century and adopted by a new generation of young people who find it useful for a wide variety of purposes, can only be clearly interpreted by analyzing the discursive context in which it appears. It is not merely students who keep it in play; "cool" and "really cool" made frequent appearances in transcripts from at least one of the principal investigators. One speculative possibility is that "cool" might serve to generalize and diffuse emotional experience for the speaker; it is perhaps a convenient and socially sanctioned stand-in for more precise emotional expression.

As a class of motivation, "cool" aspects of practice might be seen to contrast with what we have coded as a pragmatic focus in science practice. We found multiple examples of this pragmatic focus as well, so our claim is not that researchers were *merely* motivated by what is cool or exciting, only that "cool" science is one seemingly important source of motivation. "Cool" and "exciting" for the engineering scientist often go hand in hand with a potential for applications, as in the following account the PI of Lab D gives of his entry into neuroengineering; it is at once the fact that the material is theoretically exciting and affords useful applications that seems most closely associated with the affective tone of the narrative:

> So that was the first time I've ever been exposed to, even though it is fairly abstract, a model of the brain and it being **useful for something other than for something just to understand how brains work**, um, to actually use it for engineering tasks or pattern recognition or whatever. So **that was extremely exciting to me** ... and I made a road trip up to go and hear him give a talk and got even more inspired. And at that point I had decided, ok, neuroscience is what I want to do. I want to be a neuroscientist [2002–12–11-D-i-D6].

This same researcher also projects his anticipated life satisfaction on practical *accomplishment: "The thing that will make me happy when I go to my grave ... having done ... would to have accomplished tangible improvements in the happiness of people you know"* [2002–12–11-D-i-D6].

ANTHROPOMORPHIZING EXPRESSIONS: HAPPY CELLS

The class of emotional expression we found especially interesting in interview transcripts is the attribution of emotional states to the objects and artifacts central to the sense-making practices in the laboratories. The most

striking and theoretically important example of this practice is the attribution of happiness to cells in Lab D:

> I: You mentioned "cell density." Why is that important?
> D4: *Cell density is important, because for one cells survive more if they, if they're connected to each other.* **A lone cell by itself is not very happy** [2003–03–20-D-i-D4].

> D6: *So we also have microscopes in here that as you can see, they're always closed up in these boxes that we can keep warm in there, and we can modulate the atmosphere* **so the cells are staying happy** [2007–02–14-D-i-D6].

> D24: *This is problem solving on a whole new level because it's like how do we build a device that you put the microscope in that's gonna keep the humidity and the temperature in and we can deliver this to it and* **keep our cells happy** [2003–09–16-D-i-D24].

> D24: *There's neurons in a dish and* **just count them and keep them happy** *and hope they don't get sick and die.*
> I: So these MEAs, did you know anything about them before you came in the lab?
> D24: *Um, not like specifics. I knew neurons were in them and I knew electrodes monitored them and that sort of thing* [2003–09–16-D-i-D24].

> D28: *Imaging with cells is also nontrivial because there is, you know* **they're not happy** *being zapped with laser beams* [2005–03–15-D-i-D28].

Happiness is occasionally extended to the dish as a collective:

> D4: *So yeah there're, there are a couple of noise sources in there. . . . So yeah, that's the deal, it's pretty good, works very well;* **today the dish is happy** 2003–09–15-D-i-D4].

And to the cell culture:

> I: So uh . . . I don't think I understand conditioning.
> D15: *So conditioning basically means. . . . You'd basically culture an entire flask with brain cells . . . and put the media in there and let them grow in there for a week and then take the media off and use that media to feed other cells.*
> I: OK, and what does that gain you?
> D15: *Because the glia in the culture . . . maybe the neurons too . . . they're producing factors that* **keep the culture happy** [2004–10–28-D-i-D15].

Researchers across levels of expertise engage in this practice of attributing happiness to cells, from the PI to D24, an undergraduate who admits

little previous experience with or knowledge of MEAs. Therefore the attribution of happiness seems to indicate a local cultural practice (i.e., local to Lab D, the field of neuroengineering, or possibly biomedical engineering), just as the expression of "cool" reflects wider cultural and perhaps generational patterns.

Even more intriguing is the following segment of an interview with D4 that reveals a *normative* component to the concern with happy cells (neurons). This passage is instructive for two related reasons: (1) It provides more clarity on the meaning of "happy" in the context of Lab D (as we shall note, the meaning seems slightly different within Lab A), and (2) it reflects an expectation for researchers to keep the cells happy and, even more revealingly, to care about keeping them happy. The context is an early interview with D4, a postdoctoral researcher; it reflects the ethnographer's effort to better understand the basic research practices important to the lab in addition to D4's particular interests:

> I: So tell me about these neurons, because you guys are always talking about neurons in the lab.
>
> D4: *What about neurons?*
>
> I: Well, yesterday for example you killed a whole batch of neurons, you know when you were trying to get the MEA to work? And uh, remember Grad A came in and you guys, and um, what was the girl's name? She was so upset when the neurons were killed.
>
> D4: *Right, Grad B.*
>
> I: Yeah, and you guys talk about the neurons with the undergrads all the time because they're culturing the neurons, so, why do you guys care so much about neurons?
>
> D4: *If you don't essentially care about the nodes in the network, . . . **you gotta care about them!***
>
> I: But there's so many of them!
>
> D4: *Right there's so many of them, but that doesn't mean that you want to. . . . So they make up the network, **each of them has a part to play**, in the network property, so you want to keep as many as you can. You know, because they make up essentially, as I said, they make some basic rules for the way the network works, **so you want to keep them happy** [2003–04–09-D-i-D4].*

The implication here is that it is not only the happiness of the cells as a collective that is important but even the happiness of individual cells, given that each of them contributes in a particular way to the functioning of the whole. In a later interview with another interviewer, D4 provides more clues as to what is meant by happy cells:

i: So, you know, I see these sparks here [pointing to the monitor], what, can you give me a sense when you see the activity as it flows here, what, how would you describe that?

d4: *So what we see is actually summation of the activity of all the cells around it and-uh so, and this flow of activity is basically because of the connectivity of the network. . . . But then there are some cells which are always active like this guy here, and that guy there.* **They are just happy firing all over on their own.**

i: Which one, on the grid 70?

d4: *Yeah, on the grid 70. This* **seems to be happy all the time,** *you know,* **it's happy, it's happy.** *I can't really say. May be they're two cells connected to each other, they go bing-bing-bing all the time, I don't know. So there are some cells which are very active, most of the time. So, this is a* **pretty active network** *because I get activity almost like on 80% of recording, recordable channels, which is, which is very good* [2003–09–15-D-i-D4].

D4's comments in the context of this interview help clarify that "happy" cells are not simply alive but that they are also actively forming connections.

This passage in particular invites analysis in terms of the social dimension of emotion – not merely the contribution of cultural and linguistic practices to emotional expression, but more strongly the idea that emotions themselves constitute interactions or *transactions*. Transactional models of emotion emphasize the dynamic interrelations of person and environment or person and situation in the production of an emotional response. Lazarus and Folkman have provided a clear account of the transactional model:

> What does it mean to speak of relationship or transaction? The essential point is that we cannot understand the emotional life solely from the standpoint of the person *or* the environment per se. We need a language of relationships in which the two basic subsystems, person and environment, are conjoined and considered at a new level of analysis. By this we mean that in their relationship their independent identities are lost in favor of a new condition or state . . . the terms relationship and transaction are for all intents and purposes interchangeable, although transaction emphasizes more the dynamic interplay of the variables, whereas relationship emphasizes their confluence and organic unity (1987, pp. 142–143).

In this account, a range of internal and external factors (variables) interchangeably serve as antecedents, mediators, moderators, or consequences of emotion. As Lazarus (2006) later put it more succinctly, "[E]ach emotion has a different story to tell about an ongoing relationship with the environment" (p. 34).

Because of their core emphasis on interrelationship, Lazarus and Folkman (1987) linked the transactional theory to earlier systems models that were gaining momentum in social psychology. Various strands of research have provided evidence supporting the view of emotions as forms of transaction. For example, Fridlund (1994) and Fridlund and Russell (2006) analyzed the function of facial expressions and argued that they should be regarded as "social tools" that shape social interactions. Parkinson, Fischer, and Manstead (2005) argued for emotion's function in structuring social interactions in both dyadic and larger group interrelations. More recently, Griffiths and Scarantino (2008) drew from the literature on emotion as transactional in offering what they called a situated theory of emotion, wherein "[b]ehaviors which have traditionally been viewed as involuntary expressions of the organism's psychological state are instead viewed as signals designed to influence the behavior of other organisms, or as strategic 'moves' in an ongoing transaction between organisms" (pp. 439–440).

More broadly, contemporary transactional theories of emotion find precedent in Dewey's treatment of emotion in the later period of his work. Within this context, "the situation" is always the focus of analysis; the dynamically integrated nature of the situation is made clear by the participation of the actor's body (including dispositional tendencies) and personal/cultural history with the environment (including the social environment – other people) in constituting and transforming each situation. Emirbayer and Goldberg (2005) have provided a helpful summary of Dewey's transactional theory of emotion:

> Actors are always implicated in relations with other actors, and emotions cannot be extricated from those relations or seen as the properties of some disengaged or disembedded subjectivity. Not the subject (or object) alone, but rather, transactions among two or more actors (or other elements of a situation) must be deemed the proper unit of analysis for the study of the emotions. As Dewey puts it, "Emotion in its ordinary sense is something called out by objects, physical and personal; it is response to an objective situation. . . . Emotion is an indication of *intimate participation*, in a more or less excited way in some scene of nature or life" (Emirbayer & Goldberg, 2005, p. 490, emphasis added).

We find the view of emotion as the dynamic interplay between person and environment and as "intimate participation" with the environment in which practice is situated to be particularly useful in our understanding of anthropomorphizing expressions in our interview transcripts. For example, the intimate participation of researcher with cells reflected in the attribution of happiness to cells is encapsulated powerfully in this transcript:

I: So how do you keep them happy?

D4: *Pfft, you keep them happy by feeding them, by taking care of them, hopefully stimulating them* [in a motherly condescending voice – note from transcriber] *and telling them to do something! I don't know what to do to make them happy. I don't know how to make them happy, that'll make **my** neurons happy* [points to her head].

I: Hah, to make your neurons happy, your brain neurons happy.

D4: *So, my experiment is very... so we're writing a paper about, about this burst suppression* [2003–04–09-D-i-D4].

This passage points to the dynamic interplay between the attributed happiness of the cells and the researcher's cognitive goals. D4 directly ties the "happiness" of her own neurons, which we interpret as meaning her own scientific thinking and problem solving, to the well-being of the cells, despite her saying that she does not know what to do to make the neurons happy. In the interview, this passage immediately follows the researcher's claim that the cells make the rules for the way the network works, thereby requiring researchers to care about their happiness. Earlier, as noted, in the same interview, D4 expresses frustration over not knowing how to proceed in relation to her own research. The connection between the happiness of the cells and her own research progress is striking.

This notion of emotion as an indication of intimate participation nicely complements other concepts with which we have been working to characterize cognitive practices in the laboratories under our investigation. The attribution of happiness to cells in Lab D reflects a broader pattern of anthropomorphizing cells and artifacts (simulation devices) evident across both laboratories. This pattern is particularly evident in relation to cells:

A11: *The cells, once they are in the constructs, will **reorganize it** and secrete a new matrix* [2002–01–14-A-i-A11].

D2: *They (cells) have different functions and they, **they talk to each other differently*** [2002–12–05-D-i-D2].

D6: *So **they're feeling and touching and probing around in their environment until they find the right partners to form a network with.** . . . They have a lot in common, but **each dish definitely has its own personality** in terms of what the electrical activity looks like when you plug it in* [12–02–2002–D-D6 (Tour)].

Elsewhere we related the anthropormorphizing practice to what we termed "cognitive partnering," an expression of working cooperatively with specific artifacts in laboratory practice.[6] We suggested that anthropomorphizing expressions reflect the researchers' attribution of forms of agency to objects and artifacts essential to their research, suggesting that researchers engage sympathetically in transactions with them toward problem-solving goals.

Determining how the transactional perspective on emotion might inform the practice of *attributing* emotion is the task we face in trying to understand the practice of describing neurons as happy, which is exhibited so vividly in Lab D. Some transactional theories of emotion emphasize that an emotional expression constitutes *a demand* for a response. This is one way in which it can be understood as a transaction that shapes or affects the future of the relationship. In projecting onto or attributing the ability to feel happiness to cells, the researcher attaches a demand to those cells, a demand to tend them and care for them in order to keep them alive and active. This implicates the normative dimension noted in relation to the interview with D4: There is a laboratory mandate to care about and for the cells. They could die if they do not receive care. Moreover, as D4 notes, happy neurons are active in forming connections with other neurons. The study of these activity patterns is central to the business of Lab D. Similarly in Lab A, although we found much fewer attributions of happiness to cells, we found frequent references to caring for or "babysitting cells." For both labs, happy cells are vitally important to the cognitive practices, the construction of knowledge, and, thus, the dissemination of knowledge.

Note that the demand associated with the attribution of happiness to cells is context specific (specific to Lab D). In Lab A, cells are also kept "happy," but in that context keeping them happy has different implications than it has in Lab D. Indeed, instead of helping cells proliferate as in Lab D, in Lab A keeping them happy is sometimes associated with controlling or limiting their number:

> I: Um, okay, so, what have you been doing lately?
> A22: *Keeping my cells alive [laughs] and um, and happy . . .*
> I: So when you say keeping the cells happy –
> A22: Oh, you know, splitting the cells, passaging them so there are not too many and then, things like that [2002–11–06-A-i-A22].

---

[6] See Nersessian, Kurz-Milcke, Newstetter, and Davies (2003); Osbeck and Nersessian (2006); and Nersessian (2009).

Despite the differences in the meaning of "happy" cells in Lab A, we noted frequent references to the need to care for and look after them, with one researcher going so far as to identify her cells as her children. We originally coded these statements as indicating identity, specifically identity as a caretaker of cells; however, there is an emotional tone to such statements that appears to serve a similar function to the attribution of happiness to cells in Lab D. In each case the emotional involvement with cells contributes to actively working to promote their well-being. It is thus central to the cognitive practices necessary to the success of the laboratory generally and to each researcher's individual problem solving.

SUMMARY

In this chapter we have identified three classes and contexts of expression having to do with emotion, motivation, or affect. We examined their function in the context of overt expressions, figurative expressions, and expressions attributing emotions to material objects, namely neurons. The analysis, though still in early stages, gives reason to consider emotional expression to be intimately and importantly tied to the cognitive achievements and social negotiations of laboratory practices, an important aspect of psychology of science. Although the qualitative methodology adopted in this study does not afford robust generalizations, our analysis has important illustrative value relevant to the burgeoning trend of recognizing the important role of affective cognition in problem solving. In line with the research of Thagard discussed earlier, our analysis demonstrates that another rich source for investigating emotion in science practice is the open-ended interview that allows examination of natural expression in the context of ongoing research. Our study illustrates the creative ways in which qualitative analysis of the narrative accounts of scientists can complement more traditional psychological methods geared toward prediction and generalization. Moreover, we demonstrate that, to identify the role of emotion in scientific discourse, there is a need to look beyond obvious statements to include figural and metaphorical expressions and attributions of emotional states to the objects and artifacts of scientific practice.

We turn now to the closely related topic of identity formation, in which we consider in particular the cognitive and emotional effects of social positioning in the laboratories.

5

The Positioning Person

INTRODUCTION: IDENTITY AND PSYCHOLOGY

In the previous chapter we examined emotion in terms of sense-making because of its intricate interrelation with cognitive practices evidenced in the laboratories. We now turn to a topic closely related to emotion: identity. Like emotion, the topic of identity is critical to any focus on the acting person, the central unit of analysis we embrace. In contemporary theory identity frequently emerges as a form of action, accomplishment, display, or performance situated within networks of meanings and practices[1]; people *identify* as or with various options. Because performances are many and varied, the term "identities" might be more accurate to the subject matter.

As was the case for emotion, this subject matter is vital to our interest in avoiding both social and cognitive reductionism in analyzing the practices of the biomedical engineering laboratory communities. Wenger, for whom identity is of central concern in *Communities of Practice* (1998), noted that "the concept of identity serves as a pivot between the social and the individual, so that each can be talked about in terms of the other. It avoids a simplistic individual-social dichotomy without doing away with the distinction ... it is the social, the cultural, the historical with a human face" (Wenger, 1998, p. 145). Even more strongly, he argued that "the formation of a community of practice is also the negotiation of identities" (p. 149). For some theorists, identities are seen as always being in process, negotiated within interactions such that one is "seen as being a certain kind of person" (Gee, 2001, p. 99; Gavey, 1989). As similarly articulated much earlier

[1] See Kiesling (2006); deFina, Schiffrin, and Bamburg (2006); and Taylor (2005).

by George Herbert Mead (1934), others have recognized that identities and communities verify and consolidate each other.[2]

Unlike emotion, however, identity is not one of the standard categories of research and theory into which the academic discipline of psychology traditionally is partitioned. There is no chapter on identity theories that students are expected to master in an introductory class, and no advanced degree in "identity studies" is recognized as a legitimate qualification by the American Psychological Association or similar accrediting organizations. Identity is discussed most comprehensively in two contexts of traditional psychology: developmental theory and social psychology. Traditional or classical developmental theories treat identity as an acquisition or as something formed or achieved through lived experience. The most widely cited representative of this approach is Erik Erikson (1959), for whom identity formation is the task of the fifth stage of development, one of two possible outcomes of a "crisis" encountered at that stage. Identity is thus a condition of healthy development within this context. It is important to note that Erikson constructed identity as something with a fixed or stable form, clarity around which provides direction in relation to the accomplishment of personal goals, commitments, and values, (e.g., in one's career path, family structure, community involvements, and political and religious beliefs). This assumption of the stability or solidity of the acquisition of identity is evident in what Erikson established as a contrast to it: role *confusion.*

In social psychology, identity has emerged as a central concept in the context of theories of intergroup relations, notably social identity theory (Tajfel, 1982; Tajfel & Turner, 1979). Moghaddam (1998) located the origin of social identity theory in the minimal group paradigm research first published in the early 1970s, which demonstrated a general tendency toward ingroup bias even when groups are formed through arbitrary categorization (Tajfel, 1970; Tajfel, Billig, Bundy, & Flament, 1971). Social identity is, then, an aspect of self-definition rooted in group membership, as well as the emotions and valuations that accompany that aspect. Social identity theory posits that persons will be motivated to maintain an adequate social identity, one that is both distinct and positive in relation to other groups. J. C. Turner (1982) later articulated an explicit distinction between personal and social identity.

---

[2] See Burke and Stets (2009); Carlone and Johnson (2007); Endedy, Goldberg, and Welsh (2005); and Stryker and Burke (2000).

Social identity theory does not present identity as reducible to individual psychological properties, but instead understands identity as at least partially produced within social processes (Brown, 2000). Several important implications follow from social identity theory and more generally from the postulation of a "collective" or "relational" self-concept in this sense. One idea that followed from this line of theory is that persons see themselves and others as representatives of groups or categories, and thus one's view of others is determined at least in part by their category membership. We discuss this idea in relation to the question of negotiating culture (i.e., cultural categories) in Chapter 6 as we examine race and gender enactments as identity formations in the laboratory. A second implication is that social identity theory calls into question the idea of a stable or solid self-concept. If at least some important aspects of the meanings one attaches to one's own existence and experience are derived from one's participation in groups, self-understanding (self sense-making) or identity is necessarily shifting and malleable with changing and sometimes contingent relational configurations. That is, the composition of any group undergoes transformations, in some cases continually, and one may identify with various groups at various times for various purposes. Identity changes as a function of these transformations. A third implication is that identity is not one solid thing – one does not have *an identity*. If one identifies with various groups or, in other words, categorizes oneself in various ways, one is negotiating multiple identities at any given time. This point is particularly relevant to the analysis undertaken in our discussion of gender in Chapter 6, as we look at the varying ways female scientists negotiate the historically inharmonious categories of "woman" and "scientist."

Because social identity theory deals with self-categorization strategies, it risks being absorbed into the wider social cognition literature now dominant in social psychology. Yet the schema theory important to social cognition does insufficient justice to the fluidity of self-categorization theorized by social identity theory. The status of a schema is to some extent fixed as it is stored; it becomes "activated" under certain conditions, which implies that it "exists" implicitly as a representation whether or not it is in activated mode. By contrast, the emphasis in social identity theory is on self-categorization as an *activity* of persons for whom maintaining a positive self-regard is contingent on forging favorable self-comparisons, wherever and however required.

Nevertheless, Tajfel's theory has been critiqued on the grounds that it does not sufficiently consider historical context and discursive enactment in terms of their effects on the type, nature, and form of intergroup relations and social identity formations (Billig, 2002; Brown, 2002). Hence the individualistic framing of social identity theory is problematic for critics; social identity remains part of the *self* concept (Tajfel, 1982). That is, social identity theory presumes that personal self and social identity can be sufficiently demarcated such that the former can be an independent causal factor in the choices related to the latter.

By contrast, many social theorists, especially those embracing critical or constructionist frameworks, investigate identity as a set of fluid and contextualized identifications and treat these identifications as produced within current situations/interactions and reflective of past meanings and ideals upheld by the wider culture (e.g., Steele, 2003). These identifications typically concern reference groups or sociological categories having to do with race, ethnicity, and gender. Thus identity is understood as mediated by certain representations that are given their sense and reference within some culture (e.g., "Northerner," "psychotherapist"). Within this framework identities are not always consciously chosen; one is identified by others from birth in ways that reflect the structure and assumptions of a culture (black, male, American, etc.). At other times, of course, identities are embraced consciously and purposefully, or we might say that identities are motivated by social considerations. The jumble of identities explicitly claimed with those implicitly appropriated through socialization complicates the process of identity formation, but infuses it with creative and transformative potential.[3]

The developmental and social psychological perspectives on identity as we have presented them seem at first glance incompatible. In one context identity is a stable possession of an individual; in the other context it is fashioned and altered through social interaction. However, it is important to avoid being dismissive about either of these perspectives because they both seem to capture important features of identity concepts. The traditional treatment of identity within developmental theory captures the continuity or recalcitrance of identity, that experience is organized and unified, and that we claim it as our own. It refers by necessity to identity's historical formation, the fact that it reflects a unique bodily perspective and unique line of experience,[4] and that this history is experienced as bearing a relation

---

[3] However, the traditional social psychological approach remains strongly tied to a framework and methodology centered on individual cognition in that it implicates internal representations, even though social in origin.

[4] This is close in meaning to Kant's notion of "synthetic unity of apperception," which he established as a precondition of experience itself: "The thought that the representations given in intuition one and all belong to me, is therefore equivalent to the thought that

to our life choices. Consider this account by a Lab D researcher of his decision to pursue neuroengineering as a career path:

> *I knew I was gonna be a researcher when I was three. I always knew I would be a scientist in my whole conscious life. Uhm and maybe that's not, I sort of feel that all babies are scientists and that somehow it gets beaten out of them at some age, as opposed to, and it was not beaten out of me. I had parents who encouraged me when I asked questions, and instead of saying "stop bothering me." Uhm, so I had a father who I think always wanted to be a scientist, he was an engineer, and he was very much a tinkerer and a builder of stuff, he liked to do sciency stuff with me. So, so uh, when I went to college, I didn't know what kind of scientist I wanted to be and I just picked biochemistry because I figured that it is generally applicable to a lot of different things and, and it uh was apparent to my friends sooner than me that it was brain science that I was most interested in [2002–12–11-D-i-D6].*

This account is typical of the kinds of stories we heard from biomedical engineers in relation to their career paths and research choices. It reflects present actions as rooted in a unique past and set of aspirations and desires. The developmental approach to identity is indispensable because it aims at the very particularity that we have claimed is essential both to an integrated psychology and to an adequate understanding of science. Particularity is evidenced both in stylistic differences among scientists, their sense of how their past experience informs their present understanding, and their passionate engagement with certain research questions and methods.

Style, desire, and engagement have been explored by psychodynamic theorists in relation to identification,[5] but rarely in relation to science. Scientists seldom speak of their subjective or personal investments in their formal reports; indeed, research is described as if subjective effects have been eliminated. Yet when scientists discuss their own practices – including their accomplishments and failures – the personal dimension emerges as critical to their theoretical commitments and discoveries.[6]

At the same time, developmental accounts tend to underestimate the extent to which the groups and representations with which we identify have developmental histories of their own (Hall, 2001). The meaning of representations (e.g. "scientist," "researcher") extends beyond what they

---

I unite them in one self-consciousness, or can at least so unite them; and although this thought is not itself the consciousness of the **synthesis** of the representations, it presupposes the possibility of that synthesis. In other words, only in so far as I can grasp the manifold of the representations in one consciousness, do I call them one and all **mine** – Section 16 of section of *Analytic of Concepts* (Kant, 1965, emphasis added).

[5] See Freud (1923/1960); Horney (1945); and Vanheule and Verhaeghe (2009).
[6] See Gilbert and Mulkey (1984); Mahoney (2004); and Mitroff (1974).

signify for D6 or any other person; they are attached to a set of values, concepts, and contingencies that are socially formed and sustained. Thus, for example, in expressing his childhood desire to be a researcher (what we might call his early identification with researchers), D6 is responding to a set of ways in which "researcher" and "scientist" have been constructed and embellished within his family and his culture. In his family this is reflected in his father's aspirations and activities and in the culture by the fact that his friends "knew" he wanted to be a brain scientist before he consciously identified as such. Although the meanings of researcher and brain scientist are in one sense genuinely his own – they are uniquely configured in his experience – there is another way in which the meanings are limited, constrained, and expanded through their appropriation and use within the larger culture.

More broadly, the two qualities of identity we have discussed earlier correspond to two seemingly incompatible facts about identity's own conceptual history:

1. The topic of identity has an illustrious philosophical pedigree by virtue of its overlapping, if loose, relation to an expansive range of philosophical and psychological categories such as self, subject, ego, consciousness, individuality, and agency. Its central relation to the very conception of "person" is also emphasized in contemporary discussions of self and subjectivity (Atkins, 2005). It is therefore a philosophical concern of foundational significance.

2. Identity is centrally implicated in what has been called a "second psychology" developing in the second half of the 20th century and later. By this is meant a psychology dominated by concern with and analysis of "actual episodes of social interaction as unfolding sequential structures of meaning" (Harré & Moghaddam, 2003, p. 3). This alternative psychology, as we might call it, draws deeply on 20th-century philosophical contributions and social theory from both Anglo American and Continental traditions, including Wittgenstein (1953), Heidegger (1927/1962), Foucault (1977), but also Vygotsky (1978), Dewey (1930), Mead (1934), Goffman (1981), Garfinkel (1967), Bourdieu 1980/1990), and others.

The great core of philosophical concepts and problems to which the concept of identity relates makes even the task of defining identity no easy charge. Indeed, the definitions offered tend to reflect theoretical commitments that risk the problems of individual or collective preeminence that Wenger has argued the concept of identity is equipped to overcome.

Conceptions of identity range from sheer memory continuity – a connection of memories "from the inside" (Shoemaker, 1984, p. 89) – to the idea
that the very notion of such continuity, like that of "person," represents "not
logical or analytic features of personhood, but rather socially instituted and
maintained norms of intelligibility" (Butler, 1990, p. 16). As Wenger noted,
construing the notion of individuality as an aspect of community practice
does not constitute a denial of individuality and does not imply that participants in the community lack private experience. It only threatens the basis on
which the individual and the collective are juxtaposed: "We cannot become
human by ourselves; hence a reified, physiologically based notion of individuality misses the interconnectedness of identity. Conversely, membership
does not determine who we are in any simple way; hence generalizations
and stereotypes miss the lived complexity of identity" (1998, p. 146).

However, even within the general framework that acknowledges the
close relation of identity and community or identity and practice, multiple
ways of articulating their reciprocity have emerged in recent scholarship.
One prominent approach with several offshoots and varieties is positioning
theory. In this chapter we explore the potential of positioning theory to
inform interpersonal dynamics and identity formation in the biomedical
engineering laboratories.

POSITIONING THEORY: OVERVIEW

Positioning theory and the concept of positioning have conceptual roots
in many places. Positioning is "the discursive process whereby people are
located in conversations as observably and subjectively coherent participants
in jointly produced storylines" (Davies & Harré, 1999, p. 37). Positioning
analysis seeks to describe changes in relational configurations effected by
various discursive strategies; that is, it concerns the social effects of speech.
In other words, our speech is affected by what others have said, and our
speech affects other people. Positioning represents the action though which
these effects are accomplished, and a (social) position can be understood as
an effect. We identify ourselves and identify others by positioning ourselves
and others with our speech. Louise Alcoff has described the bidirectional
nature of positioning and its relation to identity:

> We might, then, more insightfully define identities as positioned or
> located in lived experience in which both individuals and groups work
> to construct meaning in relation to historical experience and historical
> narratives. Given this view, one might hold that when I am identified, it

is my horizon of agency that is identified. Thus, identities are not lived as a discrete and stable set of interests, but as a site from which, one must engage in the process of meaning-making and thus from which one is open to the world (Alcoff, 2006, pp. 42–43).

That positioning theory is likewise concerned with what has been traditionally labeled "cognitive" and "epistemic" is similarly emphasized in recent work: "Positioning theory is a contribution to the cognitive psychology of social action. It is concerned with revealing the explicit and implicit patterns of reasoning that are realized in the ways that people act toward others (Harré, Moghaddam, Cairnie, Rothbart, & Sabat, 2009, p. 5).

Both interactive positioning, by which one's speech positions another, and reflexive positioning, "by which one intentionally or unintentionally positions oneself" (Moghaddam, 1999, p. 75), are recognized in the form of positioning analysis we engage here. Although the interview is of course an interaction and thus includes interactive positioning, we are particularly interested in ways researchers position themselves in their accounts of their experience in the laboratories.

Precursors of positioning theory and analysis are as varied as psychoanalytic object relations theory (e.g., Fairbairn, 1954) and Vygotsky's theory of cognitive development (1978), with the sociological analysis of Erving Goffman (1981) the most important and direct line of influence. Even Adler's discussion of birth order and its effects on family dynamics and personality formation is an analysis of positioning within the family, although Adler's conception of position is more static than found in more contemporary formulations of positioning (Adler, 1927/1992). With the exception of Adler, a point of cohesion among these varied approaches is an emphasis on sociality, specifically *language* as the medium of identity production. Positions are "patterns of belief" or meaning structures as distributed among "members of a relatively coherent speech community" (Harré & Moghaddam, 2003, p. 4). Some researchers place greater emphasis on the local effects that are generated within a given conversation (Ribeiro, 2006; Wetherell, 1998). However, within the general frame of positioning theory, discourse is a form of activity, and the focus of analysis is the tacit though discernible rules (constraints), affordances, and negotiations (transactions) that shape, confine, or enable human practices within the context analyzed.

Position implies location, and indeed "people are seen as locations for social acts" (van Langenhove & Harré, 1999, p. 14). However, because discourse is in flux, the concept of position is more fluid than that of the social *role*. Role implies a static location in a social order; in contrast, positions are always dynamically shifting and renegotiated, both in their first-order (tacit) and second-order (intentional or conscious) forms. Positioning then "can be seen as a dynamic alternative to the static conception of role" (p. 14). Because positioning is constantly shifting, different positions can be identified for a single speaker in the context of a single interview, which adds substantial analytic power.

Two features of positions clarify the nature of their social effects. First, positions serve to establish the possibilities of action; that is, analysis focuses on what actions are "socially possible for any social actor at any moment in the flux of social life" (Harré & Moghaddam, 2003, pp. 4–5). A second, related point is that a position "can be looked at as a loose set of *rights and duties* that limit the possibilities of action" (p. 5). It "may also include prohibitions or denials of access to some of the local repertoire of meaningful acts" (p. 6). Positioning theory therefore emphasizes the "subtly varying presuppositions as to right of access to the local repertoire of acceptable conduct" as well as "presuppositions as to the distribution of duties to perform the necessary action" associated with various positions (p. 4). Hence different positions are understood to *make possible* different forms of practice and to *require* some forms of practice: Positions serve to establish the possibilities for action, broadly defined. Positioning is thus closely tied to identity formation *and* emotion in that shifts in position effect changes in the sets of practices in which one can and should participate and feels competent or obligated to do so.

As a linguistic or conversational phenomenon, positioning typically is analyzed in person-to-person or group-to-group interactions (or person-to-group interactions). Emphasis is on the bidirectional effects of different linguistic strategies; that is, on how specific linguistic expressions as used in a particular context subtly influence and change the dynamics between persons or within a practice community. For example, a teacher's relationships with her students shifts delicately when she requests that they call her by her first name; the effects are bidirectional in that both her standing (position) in the classroom and her activities as well as the activities of the students are affected by this seemingly simple act. Rights, privileges, and duties alter for both the students and teacher: Students might feel more intimate access to the teacher, and the teacher might feel less able to exert authority. That the identity of both teachers and students in relation to one another might similarly be modified in this transaction is also easy to imagine. The point of positioning theory is that these changes in rights and duties attending discursive shifts and strategies are constantly in play through our communication. Recent work has emphasized that shifts in rights and duties inevitably implicate power dynamics as well (Louis, 2008).

To our knowledge, positioning theory has not been extended into science studies in any wide-scale fashion, although there is increasing interest in the relationship of the cultural-historical situation to identity in science education.[7] Our intent is to not only complement but also expand on efforts to apply positioning theory to science practice, which have taken two major directions. First, positioning has been invoked to account for rhetorical strategies in scientific writing and other forms of representation used in the formal justification of discovery practices. Notably, Latour (1987) addressed positioning tactics as forms of rhetoric for presenting research results – tactics include the use of numbers, pictures, graphs, other research, names, and other supportive devices, layered in such a way as to minimize gaps in an argument, especially in relation to a scientific controversy. According to Latour, positioning is the process by which numbers are "arrayed and drilled" in a text for the purpose of convincing readers of the merits of an interpretation (1987, p. 50). Van Langenhove and Harré (1999) also analyzed positioning strategies in relation to scientific writing and publication. Second, feminist critiques of science have similarly engaged the notion of subject positions to analyze power relations implicit in scientific discourse (Wilkinson & Kitzinger, 2003). However, we consider positioning theory to still be underutilized in relation to science and believe it holds great potential as an aid to interpreting the practices of scientific knowledge construction that precede and follow formal representational practices (e.g., writing, graphing), in ways that include but are not limited to gendered relations.

Our view of the research laboratory is that it constitutes a discursive (speech) community in its own right. Its rules and conventions of conduct, along with some of its terms and concepts, are locally established through interaction particular to its community. Yet each laboratory community under investigation is embedded within wider discursive communities, including the field of biomedical engineering of which it is a part, the tradition that sustains the field, and the discourse of innovation and science practice writ large. The idea of the laboratory as a discursive

---

[7] See Tobin and Roth (2007); Gee (2001); Carlone and Johnson (2007); and Brown (2004).

community is enhanced by considering its operations as a *storyline*. As Harré and Moghaddam defined it, a storyline is a

> working hypothesis about the principles or conventions that are being followed in the unfolding of the episode that is being studied. Such a story line might be "David and Goliath" or "Doctor and patient." These titles sum up what is to be expected in the episode being studied and comprise the conventions under which to make sense of the events that have been recorded and to express them in a narrative.... Each story line incorporates positions that relate the participants in a definite way (Harré & Moghaddam, 2003, p. 9).

The overarching storyline of each laboratory – its ostensible purpose and research specialty along with the conventions that structure engineering science research more generally – is interwoven with a series of smaller story-lines representing the particular projects and interactions of the researchers therein. To some extent these stories are "written" by the conventions of academic rank and role: principal investigator, postdoc, undergraduate student. However, the laboratories under our investigation are marked by an unusual distribution of expertise that cuts across roles. Because biomedical engineering is an interdisciplinary field, there are distinct differences in the academic preparation and skill repertoire with which researchers begin as laboratory participants. Moreover, the "frontier" nature of research in each of the laboratories we study results in an atmosphere in which all the researchers occasionally feel that they are operating without a clear sense of direction. The interdisciplinary culture and the cutting-edge, exploratory problem solving combine to set up an intricate set of storylines that continuously position participants in complex and shifting ways. Even though the objective of the educational program in which most of the researchers are participating is to create individuals who are hybrid "biomedical engineers," we have noticed that researchers position themselves within this complexity in a variety of ways, with discursive strategies that anchor their work and identities more strongly in the normative frameworks of either engineering or biology and in science practice more generally. Statements made in interviews reveal these positions, the social and, as we argue, epistemic effects of which can be analyzed and compared. The interview itself is an additional storyline that interfaces with the storylines the participants are relating in the interview.

In this chapter we illustrate the usefulness of positioning theory by examining examples of discursive strategies enacted in the interviews. We analyze them in terms of rights and duties, both cognitive and social, that constrain and enable forms of sense-making. We also explore relations between positioning and indications of emotional or affective experience. As in the previous chapter, we organize selected examples into three categories to emphasize the qualitatively different ways in which positioning can be identified in laboratory practice. In this case we begin with examples in which researchers make explicit reference to professional or disciplinary affiliation, and we discuss their implications in terms of positioning theory by considering the rights and duties associated with the linguistic strategies in evidence. Second, we examine more subtle examples of positioning that relate closely to cognitive activities and knowledge construction – with ways in which ideas or methods are justified, extended, or modified in relation to the work of other researchers or scholars. Third, we extend the analysis of positioning to the realm of researchers' relationships with the artifacts and objects central to their practice, in this case living cells. This broad division into three categories of positioning is intended to distinguish three forms, levels, or dimensions of identity negotiation in laboratory practice. In addition to reflecting the categories of coding that emerged from our grounded analysis of interviews, the organizational scheme adopted for this chapter is also intended to parallel the division adopted in relation to emotion/affect themes in the previous chapter: explicit examples, figurative examples, and examples that extend our concepts to relations with laboratory objects and artifacts. Interview text examples coded as involving positioning similarly display the highly integrated nature of social, cognitive, and affective aspects of the "discursive acts" of researchers recorded as interview material.

Positioning in Relation to Professional and Academic Identity

We first examine a class of passages that express explicit identification with a disciplinary tradition or with some configuration of disciplinary traditions. Generally these passages entail alignments with either biology or engineering, but in some instances we see researchers aligning with the ways in which biology and engineering are combined in biomedical engineering or in bioengineering (these are distinguished). That is, researchers in both laboratories position themselves during the interviews as either engineers or biologists or as a living blend of the two specialties. In the context of the interviews these positioning strategies serve to define a set of procedural and conceptual rights and duties in accordance with the disciplinary identification or the combination of identifications.

As in any interdisciplinary culture, there are negotiations to be made as researchers enter the laboratory and find the place that best fits their own disciplinary background and skill set in their transition to an inter-disciplinary researcher. At a most basic level we would expect to find some division of labor in line with these differences in expertise. As an example, here A11 characterizes the differences between biology and engineering as entailing a division in both focus and task in Lab A:

> I: Now would you say this lab is primarily a biology-driven lab? Where does biology fit in?
>
> A11: *Fit in? It's in the sense that we are interested in pathways but just not chemical pathways.... And it's definitely biologically oriented in the sense that we're doing a lot of cell work, a lot of animal work. But I would say it's... I guess on my end it's not as biologically oriented. I guess you could say it's split because we have* **some people that do mechanical testing of pumps** *or just looking at micro C-T or figuring out better ways to use micro C-T. That's about half of our lab. The* **other half does lots of cell culture.** **So I consider myself in the middle of these two** [2002–03–22-A-A11].

Of interest here is not only A11's perception of a labor division between those who "*do mechanical testing*" and "*figuring out better ways to use micro C-T*" (engineers) and those who spend time engaged in culturing cells (biologists) but also A11's own (unprompted) positioning of himself as "between" these. That is, the interviewer's question concerned the role of biology in the lab, and A11's answer includes a statement about identity: his own place in relation to the disciplinary partition he describes.

Other researchers express the distinctions between engineering and biology less clearly in terms of focus (on a mechanical tester or on cells) or task (testing or culturing). Rather, for them what seems to differentiate biology and engineering are differences in norms and scientific *values*. These values implicate methods of problem solving and thus concern specific procedures, but these researchers' emphasis is not on the task so much as the approach to science characteristic of engineering or biology. By aligning with one side or another, these differences become matters of identity.

For example, A10 characterizes engineering in terms of a high degree of experimental control, in part through having well-defined variables:

> I: So is there a force in vivo that you're trying to simulate with this?
>
> A10: *Right. So the blood vessel is a complex biaxial strain.... But, it's impor-*tant as an engineer to really define your variables or your parameters, *or whatever.... If I just stretched the cells... I would be able to say, you know, my cells did x, in my device, but you know, I don't want to make it device dependent. You know, I want to say* [to address/inform] *the strain pattern.*

I: So you think that your engineering background may be motivating your . . .

A10: *Oh yeah!*

I: . . . desire to control that?

A10: *Oh yeah! Oh yeah – a lot of people will just want to stretch them, and see what happens when they're stretched . . . a lot of people will extract that kind of stuff, prematurely* [2002–07–05-A-i-A10].

In this example, the higher degree of control that A10 associates with engineering also enables more robust inferences and scientifically relevant generalizations (i.e., that relate to the strain pattern of the blood vessel rather than the specific device used to test [stretch] the cells).

In a later interview, A10 echoes the idea that engineers are concerned with high levels of experimental control to achieve the overarching goal of approximating a blood vessel's environment:

I: So why don't you briefly tell me what the flow loop is, and what it's for, in your own words.

A10: *So we use the flow loop as a first-order approximation of a blood vessel environment, is like, in that, as the blood flows over the lumen, the endothelial cells experience a shear stress.* **Well, as engineers, we try to emulate that environment. But we also try to eliminate as many extraneous variables as possible, so we can focus on the effect of one. Or perhaps two, such that our conclusions can be drawn from change in only one variable.** *So, we've come up with this flow loop as a way to impose a **very well-defined shear stress** across a very large population of cells such that their aggregate response will be due to that well-defined shear stress* [2002–07–09-A-i-A10].

A10 reiterates this emphasis on the engineer's superior control by way of contrast with "the biologist's perspective" and with ways of working that seem to be more popular with some general class of other "people" who would appear by association to be biologists:

I: But it's interesting you kind of just talked about the, an optimal situation, if you could reproduce . . . you know, or simulate a part of, a real part of the body where it does have a pulsing, like, a heart, and you have blood flow, that.

A10: ***People*** *do that* [implication – just not us].

I: They do?

A10: *Well, they just look at normal people. You know, they stick a couple catheters down there and see what's going on one side or the other, or they'll do stuff where they'll take a big money clip, and they'll clamp it over a rat aorta or something, which will cause a stenosis. . . . You know, so you can do stuff like that but um, the difficult thing is, you know, what is doing that?*

> *The shear stress? Or what if putting the stenosis causes a large growth factor of x amount?*
>
> I: You just have more control with the flow loop.
>
> A10: *Exactly.* **It's well defined, you can change one thing and therefore whatever happens it's because of that one thing.** *No matter what. . . . Whereas like a biologist's perspective, would be, you know, let's see what's happening, but the problem is the conclusion. . . . a lot of times biologists will just try to draw conclusions.*
>
> I: Like a causal model, or . . .
>
> A10: *Right.*
>
> I: And that's problematic . . .
>
> A10: **Very** *problematic.*
>
> I: Why?
>
> A10: *Because you* **do not have a firm control over all your variables.** *And you have so many variables, they all could be changing all at once, there's no way to know, all you have is a snapshot here, a snapshot here. . . . So, you know, I mean, so many of those things are, it's so difficult to ascertain which is why we do this, you know, basic kind of research. This is a lot slower. And I think that's* **why people don't like it.**
>
> I: And so are – I mean, the biologists are, are content with the kind of systemic cause and effect that they describe?
>
> A10: *They* **can get away with it in their circles.** *Engineers, you know, don't really like that too much.*
>
> I1: So do the biologists make progress with that method?
>
> A10: *Well, I mean, it's kind of like paleontology versus microbiology. Paleontology you see a bigger skull and a little skull, and you say this came from that. Microbiology* [you say] *these genes are totally different. . . .* **But a biologist would say this is this.** *You know, the really,* **engineering mindset** *would be let's look at genes, and let's see if can we actually draw this kind of relationship.*

In these passages A10's efforts to align with engineering practices and mindset and to distance from those of biology position him as a more rigorous scientist; they establish epistemic rights and duties consistent with that position. Although he does not explicitly criticize biologists, he claims a more rigorous method and solid logical foundation for his own work by identifying with engineers and claiming that engineers are more systematic in drawing experimental conclusions on the basis of eliminative induction.

A11, a PhD student, also expresses the conviction that engineering offers cognitive advantages. Of interest here is that the interviewer's question concerns *strengths*, and A11 answers in terms of disciplinary identification:

I: So what are your strengths coming into the PhD program?

A11: *Well, I did a bioengineering... biomedical engineering background, so I had an engineering sort of discipline with a material science background so I would say kind of the materials side of things.* **Understanding more of the mathematics and things like that...**

It is clear from the context that A11's implied contrast is biology:

I: So what are your plans for the future?

A11: *I would like to work in some sort of medical device sort of uh project. So similar to what I did either on my internship or in a different uh company and things like that. And also working more in an engineering sort of discipline.*

I: Rather than biology?

A11: *Rather than biology.*

I: Why?

A11: *Uh* **probably some of it has to do with the fact that the engineering aspect was more successful than the biology** *was, but I also think that the questions that you have* **in biology you really can't answer a lot of them very easily because no one really knows what's going on.** *Whereas questions with engineering you may make very difficult equations and very difficult to solve, but you can at least approximate them or reduce the problem down to something that's simpler and may not be appropriate for every aspect of what you are looking at but at least you can kind of address it on some level whereas biology is like a big black box. You don't really know what's going on because everything interrelates to everything else.*

In this passage A11 expresses his sense that the engineering side of biomedical engineering affords greater scientific efficacy than does biology. Engineering questions can be addressed more easily because of the methods employed in engineering, in this case involving engineered approximations of bodily environments.

The advantages ("rights") offered by biology emerge later in A11's account of biomedical engineering, even as he acknowledges a widespread view of engineering as a "hard" science and biology as a "soft" one:

For some reason a lot of engineering faculty think that engineering is a more... I guess hard science as opposed to biology, which is still relatively soft. They're afraid that if just by being called bioengineering it's not really a true engineering. They're not doing like a true science work, which isn't really true. But I think because of that, not only the faculty but students as well wanna make sure they don't lose touch with the engineering in their projects. So I think most people come in, especially here they come in from a more traditional engineering background, ME or ChemE, sort of want to make sure that they keep those in their projects. So they tend to start out trying to do a true engineering project. **Then they realize that with tissue engineering they can't really like answer all the questions, or all the important questions with just engineering so then they end up switching over to biology sorts of approaches and end up doing more biology near the end.**

The idea here that engineering is something with which students and faculty do not want to "lose touch" underscores its position as a more rigorous approach to scientific problem solving, with "hard" likely referring to the superior control and quantitative manipulation of the features of biological phenomena afforded by engineering. However, A11 also suggests in this passage that at some point the affordances of engineering run out and control must be exchanged for the broader set of questions and methods that biology can provide.

This view of biology as offering a more open-ended, theoretical approach to problem solving is presented by A14, who expresses the most explicit identification with biology of anyone in Lab A:

I: From what X has gathered you are kind of the only biologist in the lab.
A14: *Essentially, yeah.*
I: We had kind of developed a kind of description of biologists and engineers from things that people have said [in the lab]. And I was curious about your response to this.
A14: *Even within biology you have some that are more quantitative and more descriptive so this is very generalized – but yeah, I say that biologists may build on formulas and theories you know, things that already have been quantified as far as the materials. Now when you say do not, do not invent [pause] I think there's innovation in biology so I wouldn't discount that. It's totally that the work goes on in a vacuum. You see the thing about biology is that biology is really a part of life and even though engineers may not realize it, due to training and such, biology is the thing that . . . I mean you . . . I want to say it's almost innate.*

Here A14 seems to be portraying biology as a natural, accessible realm, drawing what is perhaps an implicit contrast with the artificial constraints of engineering that enable tighter control. For A14, biology and engineering represent contrasting approaches to inquiry. He characterizes these fields as two *perspectives*; in the following passage he explains how the difference plays out in relation to a specific example, in which an engineering student is experiencing a series of frustrations in his efforts to culture cells:

I can think of one specific example – the perfect example of engineering and biology. I was working in a mechanical engineering laboratory and a student *was trying to grow some cells. . . . The cells after a couple of days would shrivel up and start to die. . . . I used the same media, used the same temperature, put them in the same incubator, and they looked at measuring things. They [the student] looked at both incubators – the temperature was the same. The components and media were the same, [but the cells] aren't growing – just aren't doing what they are supposed to do. So once again maybe from an engineering mind they looked at, you know, measuring things. Everything was the same, so I said, "Well, lets look a little bit outside of the box." . . . I asked the question what about the micro environment of the chamber – the only thing that's different in the system is the actual chamber. . . . His device was something that he created and built based on the mechanical properties. But in the design process he did not take into account that maybe some of the materials used to build his device were toxic. . . . So once again it is something of trying to **look at the whole picture instead of being focused on one thing** – getting the cells to grow and stick. And so once again **two different perspectives** [2001–A-i-A14].*

Although the ostensible point of A14's storyline is a description of biology and engineering as offering alternative perspectives on science, there is a clear positioning of biology as an ultimate authority because of its more holistic and systemic focus. The authority thus conferred on biology in turn positions A14, the sole biologist, as an arbiter of natural wisdom, as underscored by his suggestion that engineers come to him for counsel: "*What happens in some instances is* **they end up coming back to me and saying well I need a little bit more advice**" [2001-Ai–A14].

By contrast, a researcher in Lab D positions biology as concerned with identifying causal mechanisms, but as unequipped to provide a theoretical account of their workings. This contrast is drawn as D7 tells the interviewer the story of how he joined Lab D:

D7: *I started searching for people who are doing neuroscience work like mine, so then I can apply my physics there, and model it from a physics point of view.*

I: Tell me more about what it means to model what people are doing here from a physics point of view.

D7: *What biologists do, they take some data and analyze that and say, "OK, this means that." But,* **how** *it is, like, hmmm. For example, OK, this is some spike pattern going on. And this is running, and this is being recorded from this, something in vivo, and this is the spike pattern.* **And since every time it moves the spike pattern is coming up, so this means that this is causing this or this or this,** *or this is a result of this behavior.* **That's what biologists do. Now I will ask why, how it is happening. That is physics** [2002–10-24-D-i-D7].

These examples help illustrate the difficulty of characterizing the precise nature of disciplinary divisions within the culture of the laboratories as expressed in the identifications assumed by researchers. Disciplinary configurations are made even more intricate by the complex ways in which dual identifications are enacted by some researchers through discursive positioning. We asserted earlier that it is important to consider both the relatively stable developmental trajectory of identity and the fact that social conventions and current interactions provide the conditions through which this identity can be claimed as one's own. We see the relevance of both of these considerations in the disciplinary identifications of biomedical engineers in our two research laboratories. Researchers are responding to the traditional ways in which "engineer" and "biologist" have been identified, and they are struggling to understand what it means to be a hybrid "biological engineer," a category with a less venerable and established history than those of engineering and biology. At the same time, researchers bring their own interpretations to their ways of positioning biology and engineering and biomedical engineering.

For example, after noting that his background is in biochemistry, D6 says that he is *using* neurobiology and is *interested* in cognitive science and neural modeling. He then positions his interests to constitute a kind of divided subject:

> *My background is, uh, in chemistry actually, but, I'm, I uh, I'm using neurobiology and interested in all sorts of cognitive science issues and neural modeling, that sort of thing.* **The biologist in me** *is interested in learning and memory, and how it is that we learn and what changes in our brain when we learn something. And* **the engineer in me** *is interested in taking some of those ideas and applying them to computing, figuring out completely new ways of computing that currently aren't being used in any of our human made artifacts right now* [2002–12–02-D-i-D6].

Note that in positioning himself as dually motivated by biology and engineering, D6 asserts two quite divergent sets of rights that work together to equip him for the work of Lab D. Identification as a biologist drives his questions; identity as an engineer enables the computing applications that result in the innovations he can claim. Both identities are empowering but in different ways.

A similarly mixed identity is expressed by D4, but with a less sanguine assessment of its benefits:

> *So I come from a total electronics, engineering background core: no biology. And then I kinda did a masters in double-e* [electrical engineering]

interdisciplinary biomedical engineering. So, then I started doing, well I took courses in the biomed department, somehow managed to get an inter-disciplinary masters degree in EE and biomedical, but, if people ask me in biomedical, I'm like . . . "So, What's your masters?" I say, "biomedical engineering." And people ask me in engineering, I say "EE." So . . . I can be both ways. My certificate actually says interdisciplinary. . . . BME requires both because of the engineering background – engineering skills as well as biology. It's not one little thing. . . . I tell my friends, "Well, we're not BMEs" "What is BME?" "Well you're not really an engineer, because the engineers don't like you any more, they think you're traitors" . . . Yeah! They're like, "You're not engineers, you're not electrical engineers." Because I am not doing core electrical engineering, OK. The biologists don't think that I am a real biologist because I know kind of biology, so . . . So, when we're nei-ther engineers nor biologists. We're just BMEs. [Laughs] That's how I put it. [Laughs] We're nowhere. We're somewhere, but right now we don't know anything. We don't know the biology, we don't know the engineering. So, we're somewhere in the middle [2003–03–20-D-i-D4].

Despite its humor, this account expresses inherent drawbacks associated with her own identifications as an interdisciplinary researcher and the prob-lems they create for knowledge construction in Lab D and in biomedical engineering more generally. The claim that BME *requires* both engineering skills and an understanding of biology can be analyzed from the point of view of positioning theory as a statement about the *duties required by* posi-tioning the field as an interdiscipline and positioning oneself as an inter-disciplinary researcher. There are cognitive requirements, and they have emotional corollaries that are not always easily managed and that include not feeling entirely secure about the knowledge foundations supporting new problem solving. Nevertheless, D4 also recognizes that the very confusion invited by mixed identity also serves to open inquiry: "*I think it's the way the fields are going. It's kind of nice to get a blend of both. . . . I think it broadens your outlook on things. Generally I think it's interesting.*" Thus the opening of inquiry is a *right* afforded by the positioning of oneself as negotiating two disciplinary identifications.

A26, who holds a PhD in bioengineering, expresses a similar sentiment in relation to the interdisciplinary nature of bioengineering:

I: So do you, do you consider yourself more biology or more engineering? Or kind of right down the middle?

A26: *So well, I came here to get more biology. The slight problem in my opinion of bioengineering – the way I did my career doing a bioengineering under-graduate and bioengineering graduate degree, I would change if I had the*

option, knowing where I would be now, which is I would do a more clas-
sical engineering undergrad degree.... Um, bioengineering undergraduate
degree, especially when I did it, which was very early on, there were very
few bioengineering undergraduate programs – is too interdisciplinary. And
so you don't learn strong fundamentals in any one engineering. So, how do I
view myself? I view myself as someone with an engineering bias, but a **weak**
engineer *if I compare myself to a chem E* [chemical engineering], *or mech*
E [mechanical engineering] *at this school. But not a biology student. So I'm*
kind of...

I: Mm hmm.

A26: **Which is a little disconcerting, but a little empowering.**

The characterization of an interdisciplinary program as both disconcerting
and empowering illustrates that mixed or complex disciplinary identifications affect emotional as well as cognitive functioning.

To summarize briefly, in this section we have considered a range of positioning strategies, specifically relating to explicit disciplinary alignments
associated with specific cognitive practices. In accordance with positioning theory, we have identified cognitive and/or epistemic rights and duties
associated with various disciplinary positions. In turn, emotional and motivational dimensions of experience appear to be closely tied to the identity
commitments and approaches to scientific problem solving that these commitments implicate.

Positioning as Warrant and Justification

We now turn to a class of discursive moves we classify as subtler cases of
positioning. By "subtle" we mean that they do not make explicit reference
to disciplinary identifications; rather, they demonstrate either alignment of
methods or ideas with those of other persons, laboratories, or traditions
or a distancing from others' methods or ideas. These expressions serve
to anchor the researcher's own problem solving or provide warrant for
a developing idea. We originally assigned a label of "latching" to capture
researchers' attempts to justify epistemic actions by means of alignment with
existing knowledge or traditions. In the case of an alignment, something is
conveyed as stable, reliable, or trustworthy, and it functions as something
with which one seeks to associate a project, method, or line of thinking.
Our observation that a parallel *distancing* maneuver, in which one moves
away from standard, established, or usual practices, serves similar functions
led us toward the higher level category of "positioning" and to turn to
positioning theory as a way to inform what we observed so frequently in the

interviews. This frequency is not surprising. On one hand, it might be said that all of science involves positioning, inasmuch as problems and methods are always situated in relation to others, and justification for conclusions are so obtained. However, positioning theory offers the analytic strategy of considering the (epistemic) rights and duties of these alignments and distancing maneuvers.

We begin with a rather typical example of the way in which a researcher made the decision to pursue a particular line of research by combining his own desire and interest ("what I wanted to do") with positioning what other researchers are doing. In this case he refers to other researchers in Lab A:

> I: So, what are the major research questions you are struggling with right now? . . . Tell me how you settled on this project.
>
> A4: *Um, it pretty much was a process of elimination and* **this was what I wanted to do**. *I didn't really want to work with endothelial cells; I didn't want to work with the flow loop. I really wanted to stay more with the chemical aspect and basically the whole mimicking and biomechanical influence on constructs to behave in the right manner. And so one thing A13 was talking about was the certain influence of mechanical stimuli on a biological marker so, I said OK, I had done a little bioengineering so let's look at biology* [aligning]. *Well, after I read, I realized it was a lot more biology. And it's very interesting but I think it's gonna be a lot more difficult than I expected because of my learning curve which is so steep* [distancing]. *And* **I see where I'm an asset here** *in the lab 'cause next week A10 wants to do western analysis* [aligns]. *So A41's gonna go to the X24 lab and he doesn't really do that anymore but I've done it* [replacing A41]. *I still have everything in running order. If they want to do northern analysis and even when telling me about RNA preparation,* **I was telling them about degradation and what you need to store, because I'm kinda going more along the biological route** [distances]. *And then I took this biochem class and* **they think I know something** *although I know nothing.* **But I think that'll be my influence as my knowledge grows** [2002–12–13-A-i-A4].

In addition to illustrating aligning and distancing moves, A4's description in this passage calls to mind George Herbert Mead's concept of the "generalized other," which he used to express the idea that the formation of identity or "self-concept" is conditional on awareness of the configuration of roles and expectations in the groups (social systems) of which one is a part. Mead used the analogy of a baseball game to express the notion that one needs to have a sense of what others are doing, what functions they are serving, to have a sense of where one belongs within the configuration, of what one's own position can be (Mead, 1934). A4 expresses his struggle to find a way in which he can have value in the lab not merely because of his unique experiences; he also struggles with how these experiences fit into the web of knowledge and expertise already in place within Lab A.

In the language of positioning theory, A4's series of aligning and distancing statements through which he positions himself as one who can have an influence in Lab A confer on him the right to view himself as a competent and valuable member of the laboratory community – "an asset" in his words. Yet there are also requirements (duties) included with the position, as he recognizes in his commitment to acquire the knowledge needed to have continued influence in the lab. That the "rights" associated with positioning are not merely cognitive but also emotional is evident in A4's account, as it is in a similar sentiment expressed by A17, an undergraduate working in Lab A:

I: What would you like to get out of this experience?

A17: *I want to be able to do something that . . . I want to be able to finish my project and have it make sense to me and feel like I actually did something which will help me write my paper and make sense and, um, for me to feel like I played a small part. What am I trying to say? That my experience played a small part in this lab and in going up and getting better* [2002–07–09-A-i-A17].

A51 similarly expresses how one can determine one's course of work through an initial set of aligning and distancing maneuvers. Note here that the interview question is phrased in terms of what A51 was trying to *understand* and that A51 answers this question by reference to the *interest of the PI*:

I: So you said this was in the first year – I'd kind of talked with PI about the – where you're going to go with your research, so tell me where your – what your research was at that point and what you were trying to look at, and **what you were trying to understand.**

A51: *OK, well when I first came in kind of the, the project was very broadly defined.* **He** *was interested in looking at different types of fibrocartilage which are found in various portions of the body, in the menisci in your knee, that's the biggest one, you also have a little disc in your jaw joint, it's a little piece of tissue, you have a piece in your wrist, actually, or intervertebral discs have portions of them that are very similar in structure. And so* **he was kind of interested in** *um, you know, looking at how can we look at and tissue engineer these fibrocartilage tissues, much like how other people in the lab are working with an articular cartilage. Um, so originally* **we were focused** *on this little disc in your jaw joint, mainly cause it's smaller. . . . So that's kind of how* **I started** [2002–05–01-A-i-A51].

The change of pronouns in this passage from "he" to "we" to "I" illustrates the intricate play of identity in even the choice of a research problem and the understanding of that problem, tasks traditionally identified as cognitive practices.

Positioning through aligning and distancing takes place not only within the social group of the laboratory but also in relation to the wider field within which one's work is situated. This form of positioning relates closely to the norms of good practice upheld by the wider community of scientists. These norms inform both the questions it is legitimate to ask in a given scientific context and the methods and techniques one is empowered (has the *right*) to use in addressing them. In the following passage, A5 demonstrates a broader form of positioning (i.e., outside the immediate context of Lab A) in defending the appropriateness and usefulness of an experimental technique:

> i: Have you ever tried to write a program that would allow you to measure something new from a machine? Or you use existing software and machines?
>
> A5: *When it comes to analysis... I have not designed my own instrument.... [laughs] That is sort of the cardinal rule of PhDs – don't do that...*
>
> i: Don't?
>
> A5: *Because if it doesn't work you have nothing... so uh [she laughs]... but I have definitely brought in experiment techniques into the lab that were not there previously... so I had to read papers and talk to companies and read protocols and figured it out... so, we do that a lot.* **But it is also nice if you can use an existing technique so you can compare it to other people, what other people are doing.... So that is sort of an advantage... people are more accepting if it is a gold standard technique.... It's kind of a balance.... You have to find something that is appropriate for your question, but you know it has got to be useful as well** [2005–02–17-A-i-A5].

Similarly, D4 justifies her model and procedure against a generalized background of "what people do usually" and the problems therein:

> **What people do usually** *is take this couple of neurons and um, stimulate or probe them to see what the activity is, then put in some kind of stimulus or inducing stimulus, whatever the inducing stimulus is, and probe again to see what's happening... did it produce a change, did it or not?* **I think that's a very static model of studying the brain** *'cause the brain doesn't really work that way.... So,* **my way of doing it,** *I'm going to have something that is, this series is going to be like a background stimulation.* **My model is**

*basically the brain has, is getting inputs all the time. **My protocol** works with getting rid of bursts. So, I'm gonna have that as a continuous background stimulation which always goes on throughout the experiment and then on it, **I'll do whatever stimulation experiments I want** [2003–02–27-D-i-D4].*

Distinguishing her model and procedure through distancing what is usual or typical can be seen here as equipping D4 with the *right* of agency, as expressed in the assertion that she will do whatever stimulation experiments she wants to do. In another interview, in explaining the origin of her lab's use of the term "barrage" to refer to the activity of the dish, D4 distances her understanding of "burst" from the common practice in the literature of using the term to refer to the activity of a single neuron. Her use of "we" and "they" suggests identity with the practices of her lab in opposition to those of the literature on bursting; this distancing affords new opportunities for the development of technical vocabulary and subsequent theorizing that need not conform with the understanding of burst phenomena in the literature at large. These are rights afforded by the shift in position she effects with the distancing strategy.

This broader positioning – that is, positioning in relation not only to the laboratory configuration but also to the efforts, successes, and standards of the wider field – also includes an *historical* dimension, connecting present thinking and procedure to previously sanctioned ideas and methods. That there is always an historical dimension to positioning within the laboratory context – that positioning includes an understanding not only of what others are doing but also of what they *have* attempted or accomplished in previous problem-solving efforts – was expressed in our group's previous characterization of the labs as evolving distributed cognitive systems and was central to the analysis undertaken by Kurz-Milcke, Nersessian, and Newstetter in their paper, *"What Has History to Do with Cognition"* (2005).

In Lab D, a researcher distances his procedure from that of "most people" and aligns his own systemic focus with that of Karl Lashley:

The difference between doing this [pointing to the dissociated culture and structured slice diagrams] and the type of two electrode preparation that most people do, is [that] there's a lot of data that will tell you that whatever the memory mechanism is that is out there, you're not going to find it between two neurons. It's actually a distributed process – distributed over a population of neurons, and you know that, you see studies where they have lesions, and you have things like recovery of function. And Karl Lashley spent ten years trying to, in the 50s or 60s – I think it was the 50s – where he would go through and teach a rat to traverse a runway, and he would find what part of the brain was coding this particular radial arm or runway. And he sectioned it up into smaller and smaller pieces isolating different parts and the rat kept performing the task, and eventually he concluded that there was no memory. He was joking. Right, but what his conclusion was, was that whatever this code is, it's distributed over a large area, and **you need to be able to study the functioning across that large area** [2002–05–01-D-i-D1].

By aligning current efforts with an historical effort widely regarded as transformative in its field, the reference to Karl Lashley serves to legitimize this researcher's way of thinking and proceeding. This positioning gives him the "right" to carry on in the intended direction because there is a well-respected historical precedent.

By distancing himself from "most people," D1 is able to position himself as superior in his approach to the problem at hand. On one level this activity is no different from the standard practice of framing one's current research question in the context of previous work, as might be found in the introductions to many research papers. Yet what is interesting here is that the context is not a formal paper but rather a casual interview with an ethnographer. There is no prescribed format for justifying one's procedures to an interviewer from another academic discipline, no clear indication that the interviewer will understand the significance of alignment with Lashley. It is tempting to read this as a transposition of an expectation from the normative framework that exists for positioning formal research productions (papers, presentations) in relation to historical precedents.

Finally, we encountered a few instances in which positioning related not only to the laboratory environment or to the field of practice but also to the wider organizational system within which biomedical engineering research is situated. In this example from an interview with A10, the "rights" associated with product development (engineered tissue) are altered by positioning research laboratories in relation to a federal agency (the Food and Drug Administration):

It's a little more difficult to prove, you know, [that] x came from y. But then again, if you do things our way, it's very unlikely that we'll develop a product very quickly. The product you do develop will be very likely to work better, and you know, um, right now **regulations, like federal regulations are such so that it's up to the government to determine um, effectiveness of your product,** *whereas,* **but now, like since last March FDA is really trying to put it on to the employer to determine that.** *You know, and instead of using mean values, like actual whole statistical groupings, because obviously you got this mean value that says "yes." Well, what's the distribution? You*

*could have this huge window of "no" that **but then they get all produced
and put into people** and you know, like, that's getting better and better,
because, you know, people are becoming more conscious of these kinds of
things* [2002–07–09-A-i-A10].

A10's remarks in this passage implicate his identity as a research scientist or
perhaps as a biomedical engineering researcher more generally, rather than
as a member of Lab A or as a participant in the field of tissue engineering. He first positions the federal government as the arbitrator of standards
of an engineered product's effectiveness and then alters the government's
position in such a way that empowers the individual laboratory. This shift
opens up a set of rights previously denied – namely the right to determine
the effectiveness of one's products. It simultaneously requires a new set of
duties. A10's regard for the potential consequences of poor research design,
as indicated by a failure to consider mean scores and ignore the pattern
of distribution, is expressed in his acknowledgment that poorly designed
products get "produced and put into people." This seemingly casual statement reflects an awareness of the ethical dimension of his research efforts,
of a larger normative framework embedding his particular problem solving.

   In this section we have examined the significance of our observation that
researchers at varying levels of expertise engage in two forms of positioning
as basic sense-making strategies: alignment with or distancing from existing
practices in the laboratory, other laboratories, biomedical engineering, and
science practice at large. This form of positioning is closest to what Latour
(1987) discussed as a matter of three choices forced on scientists and engineers by the technical texts they read: "giving up (the most likely outcome),
going along, or working again through what the author did" (p. 63). Our
analysis suggests that researchers make these same choices in relation not
only to technical texts but also to their own understanding of other practices within different levels of organization: the laboratory, the domain of
research practice, biomedical engineering, and Science. They appear to use
a "map" of practices to configure their own stance and strategies. Moreover, these positioning strategies affect not only epistemic rights but also
emotional and motivational dimensions of practice.

Positioning in Relation to Laboratory Objects and Artifacts

In the previous chapter we extended the analysis of affective expressions in
interview transcripts by considering the special case of attributing emotional
experience to laboratory objects, namely cells. As a parallel, we here extend

the analysis of positioning in the laboratory to include person-to-object or person-to-artifact relations, again focusing on relations with cells. In our framework, the constituent members of each biomedical engineering community include not merely the researchers and other laboratory staff but also the objects and artifacts of biomedical engineering practice. At a minimum the community includes living cells, which are cornerstones of the engineering practices in the laboratories under observation.

Inasmuch as "positioning" is a social psychological construct, our analysis of positioning in relation to cells effectively constitutes the inclusion of cells in the social realm. That is, we are considering cells to be part of the social realm of the researcher. We highlight two principal ways in which cells may be considered participants in the community of researchers that is the biomedical engineering laboratory: through the researchers' 1) attributions of agency or experience and 2) expressions of cooperative relationship with cells.

Attributions of Agency and Experience
In the previous chapter we examined the practice, which was especially prevalent in Lab D, of attributing happiness to cells. Such attributions are part of a larger pattern of anthropomorphizing we found to be widespread in both laboratories; anthropomorphizing was especially directed toward cells, although it also was directed to the physical simulation models the researchers designed for manipulating cells, such as the flow loop. By anthropomorphizing we mean the act of ascribing human qualities and abilities to nonhuman things. Although traditionally regarded as a "sloppy" linguistic practice given the (assumed) erroneous nature of the attribution, anthropomorphism is increasingly recognized as a cross-cultural and functionally important human practice, one designed to establish order through endowing the world with social meaning (Heberlein & Adolphs, 2004; Horowitz & Bekoff, 2007). In the laboratories, anthropomorphizing took many forms, including calling cells "guys," as in this example from A7:

> *You see in this series which is basically 7 hertz and 1 hertz, you get activity even further away from the electrode and **you can see on this guy more of the activity is localized to this guy and may be spread out when you go later on. So I am basically saying that a burst is something that goes throughout this guy*** [2004–07–01-A-i-A7].

In this interview A10 refers to cells "sitting down" and "feeling effects":

> I: All parts of the cells are adhering to it?

A10: *Yeah. So **the cell sits down like this**. . . . Like if the cell is just kind of hovering over this thing, stretching this membrane is not going to do anything. So you want to be able to look at it, and see it – the cell – like **feeling the effects of something**.*

Consider these explicit depictions of cell agency:

*Uh what we're trying to study in this lab is basically learning plasticity in neurons. And I feel pretty strongly that the only way you can have learning is when you have closed loop feedback and my, my reasoning for that is the stimulation you put in should have meaning somehow, should have some type of meaning to the neurons. And, well of course, it's really hard to say, OK, we have neurons outside the brain in this little petri dish what's meaning to them, you know. It's a difficult question. But I feel that **if the neurons can actually themselves control how they get stimulated, then that stimulation has meaning in a sense** [2003–03–25–D-i-D2].*

Similarly, D4 positions the network of cells as an agent, one responsible for her occasional bewilderment and loss of direction:

D4: *But there are a lot of questions and nobody knows how to answer them. Nor does D6 [the PI].*
I: So what do you think the causes of all your struggles are?
D4: *Uh, **the network has a mind of its own!** [2003–3–20-D-D4]*

The significance and function of these anthropomorphizing expressions vary with the context in which they were used, but the attributions of agency are most interesting and perhaps the most theoretically important. For example, D4's attribution of agency positions the network of cells as having not only a cognitive but also a seemingly emotional impact on her work.

Expressions of Cooperative Relationship with Cells
The cognitive "rights" opened to D4 by positioning the network as having its own mind are best understood by revisiting our concept of "cognitive partnering," which we used to code expressions of cooperative engagement in problem solving. Although we first used the cognitive partnering code for expressions of person-to-person interaction – mentor to apprentice or peer to peer – we began to notice in the transcripts many expressions of cooperative interaction with cells or networks of cells. Cognitive partnering codes thus overlapped with and were closely associated with anthropomorphizing expressions, especially when the attributions concerned the agency of the cell or network. Cognitive partnering can also be considered a form of

positioning, when it entails a researcher positioning the cell or network as a cooperator toward some epistemic end. We previously focused principally on what cognitive partnering affords and what kinds of cognitive practices it enables (Nersessian, 2006; Osbeck & Nersessian, 2005). What is interesting in the example from D4, however, is the quite emphatic expression of costs associated with partnership. As in any partnership, there are disappointments to manage and compromises to be made; goals are impeded as well as advanced.

Because the cells in Lab D are neurons, it is not a great stretch for researchers to speak in terms of mind and even agency, however philosophically problematic this practice might be. However, Lab A transcripts reveal similar expressions of relationship with cells. Two forms of this expression are particularly interesting: taking the perspective of the cells, and identifying as caregiver of the cells. The two forms bear an important relation to one another and to the concept of cognitive partnering.

Perspective taking, the first form, involves appeals to the point of view of a cell or artery, as seen in this interview with A10:

> I: How about the – how about the size of the chamber? Is that – is that part of the, ah, is that an approximation, or . . .
>
> A10: *Well, um, no in the sense that it doesn't really approximate. Like, most arteries we look at are going to be smaller than that surface. But **from a cell's perspective, the cell sees basically a flat surface.** You know, the curvature, is maybe one over a centimeter, where as the cell is like a micrometer. You know, like 10 micrometers in diameter. It's like ten thousandth the size. **So to the cell – it has no idea that there's actually a curve to it. . . . The cell, when it looks around, just sees a flat surface.** Just like we think the earth is flat* [2002–7–09-A-i-A10].

Perspective taking is a form of positioning in that it entails a shift of location to another point of view. What is the significance of the practice? Importantly, George Herbert Mead considered perspective taking to be a distinguishing mark of sociality and to establish the limits of community:

> The limitation of social organization is found in the inability of individuals to place themselves in the perspectives of others, to take their points of view. . . . In the field of any social science **the objective data are those experiences of the individuals in which they take the attitude of the community, i.e. in which they enter into the perspectives of the other members of the community** (Mead, 1932, p. 175).

Perspective taking thus becomes a more fundamental condition of cognitive partnering than are attributions of agency. Agency can be attributed to other natural objects (e.g., storms), but those natural objects do not enter into the participatory community of an actor until the actor has at least the *ability* to assume their perspective. Note here that it is the actor who by perspective taking defines the status of the object as community participant (cognitive partner) or not. However, it is not the case that this perspective taking implies that an actor constructs the social world or that the process is arbitrary (as in "anything goes"). As noted, Mead regarded personal identity as contingent on the ability to assume the perspective of others. If there is no community without the actor, there is equally no actor without a community.

However, it is also not the case that every object has an equal likelihood of receiving attributions of agency or experience. Pets are more frequently the target of anthropomorphisms than are rocks. Their perspective is more easily taken, and it is easier to imagine that important functions are served by taking the perspective of pets than that of rocks. The owner is more likely to be diligent in feeding an animal and tending to its toiletry needs if he or she is able to take the perspective of the hungry or uncomfortable pet. Likewise in the laboratories cells receive more attentive care when their perspective can be assumed by the researcher to whose work they are essential.

Therefore, it is especially interesting that, among the first set of interview passages we labeled as having something to do with identity, were passages we coded as "identity as caretaker of cells" before we began thinking in the broader terms of positioning theory. This passage from an "exit interview" with postdoctoral researcher A8 is the most striking example:

> I: Well I know you, you often, you often refer to your cells as "my cells," you seem like you've gotten some sort of . . . you like them!
>
> A8: **Relationship? They're my children!**
>
> I2: Do you ever think of them that way?
>
> A8: *When I was, when I was, oh yeah, when I was first being trained, uh, the woman who trained me, everyone gets trained on cell culture by someone, right so. . . . And when I was a graduate student at Z15, somebody trained me.* **She called 'em children,** *I think that's a very good analogy because* **you have to feed 'em, you have to keep em alive, you have to take care of them, you know, and they, they eat, and they get hungry,** *and . . . it's a good, pretty good analogy, it's a . . .* **I do call them mine, because . . . I think of them that way.**
>
> I2: It's hard to think of a, a . . . new piece of rubber like that . . . or something like that, right?
>
> A8: *Well then you think of it as property, but not as much as* **something you're taking care of. . . .**

In this passage A8 positions herself as the caretaker of the cells, which establishes both her *right* to "call them mine" and her *duties*, which include feeding them and keeping them alive. Because the cells she cares for are central to her success as a tissue engineer, the rights invoked through this positioning are epistemic (affording problem solving), social (enabling success in her chosen profession), and presumably emotional. Worth noting in this passage is that, as we suspected in the case of the "happy cells" in Lab D, the reference to cells as "children" in Lab A seems to be a culturally approved practice in tissue engineering. A8 ties her own practice to her training in another lab (Z15) in which the researcher who trained her in cell culturing called her cells "children." However, this is not an arbitrary cultural artifact. It is an analogy, A8 notes, that "makes sense" precisely because the cells are living things.

Material Agency

Indeed, their status as living things, or rather the fact that they are things on whose lives the biomedical engineers *depend* for successful problem solving, is the second important reason we consider cells participants in the laboratory community. That is, a cat that wanders into the laboratory and the microscopic organisms that inhabit laboratory nooks and crannies are not participants in the community in the sense we are trying to describe. Of central importance for the cells' status are the dynamic tension or *resistance* effected by their status as living objects and the fact that they are central to the work of each laboratory. They do not always behave as researchers expect or want them to do, and they can die. Thus we can draw a comparison between this "resistance" and the concept of constraints by which a "rule" is understood in positioning theory. Researchers are positioned by the actions of cells, which establish rights and duties in accordance with that position.

Earlier in this chapter we noted an association between identification as an engineer and an emphasis on the benefits of experimental control. Here a graduate student with a chemical engineering background acknowledges the loss of control and possible negative implications for her progress brought about by partnering with living objects:

A7: *I think the next point for me is my proposal and to write a proposal. I* **don't think I want to hinge all of my work on animals because that's so uncontrollable.**

I: Tell me what you mean by uncontrollable in that sense.

A7: *I mean when we are in the lab we can control the media we put in, we can control their environment completely.... But when we move into an animal model it's more physiologic, the challenge then is that it's a much more complex system. So I would like to integrate some in vitro work here in the lab with um the in vivo work so that way you are not, if one thing doesn't really work you are not stuck when you are considering finishing your thesis* [2002–06–17-A-i-A7].

More strongly, D2 blames his inability to complete experiments directly on the cells: "*That's been another problem why I haven't been able to do any experiments lately, cause the old neurons have been dying*" [2002–12–05-D-i-D2].

By contrast, the living status of cells is precisely the reason A8 finds her work as an engineer engaging and enjoyable:

I: Have you changed at all on, on the how you see the relation of biology and engineering and what you're doing or . . .

A8: *No, I don't think I've changed. I think I still see myself as an engineer that's using biological tools.*

I: Do you think . . . could you, do you enjoy working with biological uh stuff more?

A8: *Oh yes. . . . Well, **I like to work with things that are living, which is why I love the cell culture. . . . I just think it's really enjoyable working with things that are living, where you interact with them and they interact back with you.***

I: So that would be different than some other forms of engineering for instance?

A8: *Well, like, you build a device, I mean you can leave 'em for a month and come back and it's going to be exactly the same. And to me that's, I don't know, **it's not, it's not quite as live, as lively*** [2003–06–19-A-i-A8].

In both laboratories, despite the best efforts of researchers to culture and nurture living cells, cells "act" in unpredicted ways, the most radical and challenging of which is to die. The very possibility of cell death creates a dynamic, even "dialectical" tension between researcher and cells. Because the actions of cells determine what range of actions is possible for the researcher, cells and researchers thus position one another with their actions, with analyzable social, emotional, and epistemic effects.

We further illustrate the application of positioning concepts to researcher–cell relationships with a more detailed example from Lab D. This interview with an undergraduate researcher took place in the fall of 2005, shortly before he was to graduate with a degree in biomedical engineering; he had been working in Lab D since the summer of 2004. In earlier interviews he had given more enthusiastic accounts of his experience in the laboratory and his enjoyment of laboratory research. Near the start of this interview he declares his recent interest in going to medical school after graduation, a goal he had never identified before to the interviewer. Although medical school is only one possibility he is considering, he is definitely ruling out graduate school in biomedical engineering or a related research field. In the interview he ties his change of career plan to his experience with cell death in Lab D:

> D32: *Well . . . after working in the D lab I was like "I really don't want to do this." I don't want to have to do research that fails all the time and . . . have to write papers and . . . It didn't seem that fun.*
>
> I: Well, the last time I talked to you it was really fun. When did it stop being fun?
>
> D32: *I was actually going to do an experiment where we were measuring uh, measured the effective simulation range of a single electrode on an MEA, like that whole semester. I mean, I talked to you and then after that that's when D4 had all the cell death so I couldn't do anything and so I just did grunt work.*

Later in the interview he provides more detail about the events surrounding the cell death. It is clear that the cells had died despite the researchers' best efforts:

> *They were using the same protocol but with mice. . . . They'd been doing the same thing over and over again and one day they just kept having cells die. And they were like "That's kind of weird." So they'd plate cells again and after two weeks they'd die and they'd just keep on doing this and they were like "What's going on?" So then they started like in the middle of the fall last year. They basically made a big chart and they were like "Okay, we're going to test for this, this and this," and they delegated everything and they called all the people that provided chemicals and found out that some of them had bad batches and they tried a different chemical, no change. I mean, they found there was a little problem with their system. They fixed all the problems and they were still having cell death* [2004–08–04-D-i-D32].

The wide-scale cell death had shifted D32's position in relation to his work in the laboratory. Rights were lost – he viewed himself as unable to do *anything*. New duties were assumed – he now did *grunt work*. The impact of this change in rights and duties is weighty. Rights that D32 previously assumed to be unproblematically his own, such as the right to do his experiment and the right to be a full-fledged participant in the intellectual life of the laboratory, are suddenly and seemingly permanently denied. He previously had the right to propose work that would give him status in the lab. He had the right and duty to generate data for the whole lab that would advance their understanding of cell interaction over time. His plan was to make a contribution by advancing their understanding of how cells in cultures interact. However, the cell death signaled a kind of resistance or "stance." The cells deprived him of his contribution, shut down his work, and even altered his career plans. In effect, they transformed him into a "grunt worker." This is not to claim that the cells have agentive *intent*, only that they behave in ways that are not determined by the researcher and that constrain and limit learning and problem solving with their "actions."

In this sense, our notion of cell agency resembles the agency attributed to material objects by Latour in recent analyses, though actor network theory locates agency not in the objects per se but in the association of humans and nonhuman participants in knowledge construction (Latour, 2005).

The cell death in this instance also appears to be associated with new disciplinary alignments. That is, changes in *identity* through the experience of cell death are evidenced in the new appeal of health care and the assumption of a stance of wanting to help others by D32:

> *I never really considered that cause it's really health care but they can tie that to the sensory like they can really make prosthetics that are tied to the nerves and that would be really awesome for me. . . . I want to improve people with problems and that would be the perfect way to do it, I think. . . . I want to have that patient interaction, I think* [2004–08–04-D-i-D32].

This new stance would appear to have a positive spin if it did not stand out against a feeling of being ineffectual and *incomplete*:

> i: You put in a lot of time in that lab.
>
> D32: *Yeah, but I feel like I haven't accomplished anything. I feel like I just did the bare minimum. I don't feel like I did the best I could do for D4's lab, so I feel like I kind of owe it to him to do something big, or . . .*
>
> i: It sounds like it was the cell death that . . . like you were all lined up to do something, right?
>
> D32: *Yeah, that's why I feel incomplete, I guess. It's not really my fault* [2004–08–04-D-i-D32].

Also of interest is the way in which this narrative is suggestive of the network of relationships in the laboratory, which are connected in various ways to the cell death and its effects. D32 identifies the cell death as an event for someone else – *"that's when D4 had the cell death"* – yet he is affected as well. Thus D4's position in relation to the cells affects the position of D32 in relation to his own problem solving and learning. Moreover, emotional effects of the cell death are described here as broad in scale across the laboratory. D32 affirms that all the graduate students were angry (*"super pissed off"*), and that the experience was *demoralizing* for them.

Of course, the fate of other objects, artifacts, and devices might similarly hamper or enhance work in the laboratory. Equipment failure might be just as effective in ruining D32's experiment and prodding him away from a research career. However, as noted, anthropomorphisms in the interviews more commonly apply to cells, and we see researchers sometimes assume the cell's perspective for cognitive purposes. These actions confer sociality and social meaning onto the cells. In turn, cells "speak" through actions resulting from what is done to them and, in speaking, elicit a response.

SUMMARY

In this chapter we have considered possibilities for using positioning theory as a framework for organizing a set of codes emerging from our analysis that implicate identity and identifications in various ways. Our claims in relation to this task are modest. As Harré and van Langenhove cautioned, positioning theory "should not be regarded as a 'general theory' that calls for a deterministic application to several specific subject matters. It is not like gravitational theory. Rather, it is to be treated as a starting point for reflecting upon the many different aspects of social life" (1999, pp. 9–10). Positioning theory is compatible with and relevant to the integrated unit of analysis we seek in this book: Its particular strength

> is that it recognizes both the constitutive force of discourse, and in partic-
> ular of discursive practices, and at the same time recognizes that people
> are capable of exercising *choice* in relation to those practices. . . . Once
> having taken up a particular position as one's own, a person inevitably
> sees the world from the vantage point of that position and in terms
> of the particular images, metaphors, storylines and concepts which are
> made relevant within the particular discursive practice in which they are
> positioned. At least a possibility of notional choice is inevitably involved
> because there are many and contradictory discursive practices that each
> person could engage in. Among the products of discursive practices are
> important aspects of the very *persons* who engage in them (Davies &
> Harré, 1999, p. 35, emphases added).

Positioning analysis is linguistic analysis, specifically conversational analysis, and we have focused on the effects of speech as analyzed in the context of interview conversations. Our analysis is admittedly one-sided and departs from conventional procedure in that we have not examined how interviewers are positioned by the speech acts of researchers. Instead we have focused on ways in which researchers position themselves in relation to other researchers and within wider disciplinary and institutional frameworks. We thus use positioning analysis to address personal and professional identifications. However, we have also considered more subtle ways in which positioning through alignment and distancing in relation to others serves important cognitive and epistemic functions. Finally, because positioning theory is generally used in analyzing conversations, we have moved even farther afield in extending our analysis to researchers' "interactions" with cells and in proposing a view of cell *death*, even the potential for death, as analogous to a speech act in its potential to shape and constrain human action. We have illustrated this in the context of a young researcher's experience, whereby the death of cells effects a shift in position with implications for his learning, motivation, ambitions, and identity. The illustrative value of this material points to the potential for extending the social psychology of science to include laboratory objects and artifacts in their intricate relations to scientists. We thus abide by the spirit if not the letter of the law in applying positioning theory, given that "discursive practices" signify "all the ways in which people actively produce social and epistemic realities" (Davies & Harré, 1999, p. 34).

In each of the three classes of positioning we have examined here, identities are achieved through the discursive strategies employed in the context of interviews about biomedical engineering practice. These identities are not merely social because the effects of the discursive strategies are epistemic; they are not merely collective because they are bound to developmental histories and to the emotions and desires of every speaker.

6

The Person Negotiating Cultural Identities

The discussion of race and gender is relevant to various subfields of traditional or mainstream social psychology: the psychology of women, the psychology of race (or "the psychology of ethnic minorities" as it is referenced by the American Psychological Association), and the psychological study of social issues. At the same time, the central themes and basic assumptions of this chapter are in keeping with the "second" social psychology, which focuses on analyzing the personal and cultural meanings revealed in real-time episodes of interaction (Harré & Moghaddam, 2003). The centrality of race and gender to both the "first" and "second" social psychologies reflects the status of these categories culturally. Gender and race are what Styker refers to as "master statuses" (cited in Deaux, Reid, Mizrahi, & Ethier, 1995). One is seldom, if ever, without race, particularly in a society with a history of explicit racism (Alcoff, 2006; Fanon, 1952/1967). Nor can one escape gender; the intimacy of such ascriptions to the person and their social salience cannot be denied. A person is always identified as raced and gendered and classified according to the criteria available within a culture. However, rationality and science are typically viewed as inattentive to these ascriptions; race and gender should not matter to science (Keller, 2001).

Yet if they are inescapable, race and gender are brought into the research laboratory in *some* way, however removed from cultural effects science traditionally is assumed to be. How one *negotiates* these classifications, acting in or around or through them, is of central interest in understanding how sociality and culture operate in a research laboratory.

Our analysis of biomedical engineering laboratories yielded insights into ways in which science practice is inflected with activities best described as cultural negotiation, which are important to a full conception of the acting person of science. Among the questions that arise in considering race and gender are the following: How are race and gender negotiated within laboratory practices, including cognitive practices, and how are they intrinsic to learning trajectories in the laboratory? How do science practices implicate social and cultural representations, as well as personal histories and motivations that reflect negotiation of cultural identity? Everyday laboratory practices such as teaching a new lab member how to build simulation models, for example, would seem at first glance to be free of race and gender inflection. However, the acting person is always positioned by others and positions him- or herself in science practice as elsewhere; such positioning as reported and performed by women and minorities includes alignments and histories with respect to science and the myth of scientific purity. The lab community functions under the umbrella of a socio-cognitive pact that has historically gravitated toward certain meanings, bodies, and actions (Keller & Grontkowski, 1983/1996; Stepan, 1986/1996). The interaction of the individual researcher with this pact is of central importance in this analysis.

In the discussion of race, we examine the operation of the social pact of science practice as it is (re)created and sustained through interactions within a lab; such interactions position researchers in some ways that are empowering and in some ways that are exclusionary. In the section on gender we focus on the integration of identities or the conflicted coexistence of identities within specific practices such as experimentation, expressed through the particularity of the person for whom an important goal is to become a scientist. With both race and gender, we indicate the multilayered and sometimes conflicting streams of meanings through which one positions oneself as a developing and practicing scientist. Consistent with the overall goals of this text, we approach these two "specialized" areas of social psychology (race and gender) in terms of what our analysis of laboratory practices suggests about persons more generally.

The following two sections examine in "micro-moments" (Brown, 2004) the interrelation of identity as a scientist with other dimensions of identity such as gender and race. The process of "recovering identity" in science has been identified, but with cautions that identity is embedded in a complicated tangle of enactments; it does not function as an external authority that directs such practices (Endedy, Goldberg, & Welsh, 2005). We say much the same thing here: Gender and race are not "outside" the lab and cognition "inside" the lab. Rather, the two exist as an admixture of processes and positioning that infuse lab life in its myriad formations. As we discussed in introducing the notion of positioning, identity is an enactment, co-constituted within a community that is itself defined by current practices and representations while dependent on a history of practices and representations (Johnson-Bailey, 2004; Lloyd, 1993/1996).

The conception of identity as enactment is crucial to the present chapter, and we discuss enactment in terms of both social processes and social cognition. The section on race speaks to group processes as the foundational social pact undergirding social recognition and participation; the section on gender deals with intersecting trails of social representations and modes of integrating gender identity and gender difference within the lab. However, social psychology, although more open to the issues of context and enactment than many other subfields in the discipline, still often assumes an individualistic framework. Social cognitions are representations of reality "in the head" that affect social process (first impressions, fundamental attribution error, etc.). Group processes are conceived as impinging on the individual and affecting judgment and actions, as can be seen in studies on conformity, norms, or deindividuation. Our text, instead, construes acting as necessarily entailing a confluence between the social arrangement and personal expression. Any act is a "compromise formation," so to speak, between the collective culture and the individual actor. In the same way, any practice is an instantiation that draws from individual and social history as well as contemporary collective, cognitive, and material constraints. Thus the person, though conceptualized as central, is not treated as an isolated individual. One's emotions, social and collective history, and the positioning that occurs through interactions with others are all critical in the path of becoming a scientist.

In our research, race and gender often, but not always, played out differently. Although women in the study *might* have experienced the effects of sexual differences in the lab, they did not emphasize gender discrimination in their interviews. The following are a few examples of responses to queries about experiences of gender difference or discrimination. Given that Lab A was composed predominantly of women, this question was sometimes posed in terms of whether being in an all-female lab made any difference to ways of working in the lab when speaking to Lab A participants:

> *Not really. I can't really say, because I'm not one of those people who like, "Oh, if the lab's not predominantly women, I can't work" or "If it's not mixed, I can't work". That really doesn't matter too much to me.* [I: Uh huh.] *So, I don't know how to answer that* [laughs] *it really didn't* [2004–02–06-A-A3].

At other points, the interviewer simply asked about gender discrimination (next example, A7 [female]) or gender difference (second example, D17 [male]):

A7: *Um, I don't think it's happened to me, directly.* [I: Uh huh.] *Um, I've definitely seen like other cases where um, either students or colleagues have expressed that* [I: Uh huh], *you know, that was going on. Um, and I've been sensitive to that fact. Um, fortunately, for me, I haven't experienced that. Um, so –* [2004–01–30-i-A-A7].

———————

D17: *I don't know, because I love that kind of environment.* [I: Uh huh.] *Um, wherever I work, you know, the church I attend, or just when I am surrounded by friends, it's always like that and I, you know, and it's always been that way here at school, so, I mean, I just, it's something that doesn't really come up until, I mean when you're talking about school work or the lab work or your research* [I: Uh huh.] *it sort of just doesn't,* [I: Uh huh.] *you don't really see it, or it's not* [I: Right.] *it's just, but um, it, it's just interesting I guess whenever you start talking about your own culture's differences and things like that, but um, I guess it's never really something you notice on a daily basis.*

I: It's something, so you don't really notice most of the time, but does it like add any kind of thing to the lab?

D17: *Um, well I mean, each person has their own personality* [2004–07-D-D17].

In the words of one participant, gender discrimination is "easily over-turnable." As we examined interviews for the effects of gender within labs, we saw among students, at least, that the enactments of gender into the fold of their scientific activities were heterogeneous and not always consciously experienced as conflicted or problematic, even when narrative evidence might suggest some difficulties. Thus we looked at the transcripts to discover strategies of social and cognitive integration (expressed through particularity), using more actively interpretative readings. A woman scientist, especially in the earlier stages of her career, might allude to and indicate obliquely that gender issues have complicated her path in science or even enriched it, but many participants were not self-identified with their gender when thinking about their careers in science.

However, race was readily acknowledged as being a barrier to full inclusion in the community of the lab. As a hurdle to linking with other students and with faculty, race served or was enacted through social interactions with a draining effect on research and motivation. Participants seemed fully aware of this dimension, which was thematic despite a general reluctance to invoke racial discrimination as an explanation of a particular event. The following account is from a participant in a focus group on race and science (designated SP for Spencer):

> SP7: *I'm not going to lie. If something happens, I'll take an African American's side over someone else's side, and that's just I guess instinctive – I mean, there are things that come up and we both* [tape inaudible] *have stories and most of the time, I'll just be like, I'll just work it out, you know, keep doing what you're doing. Especially like if it's another grad student talking about their advisor, thinking that maybe their advisor doesn't let them do certain things, because they're Black. I'm like, you know, don't think about it that way. You've got to do your work, but in the back of my mind, I do wonder if it's a race thing.*

In our research, the sole exception to "race trumping gender" was the account of a very light-skinned African American who had been told by more than one faculty that she was "*too pretty*" to be a scientist. All other participants interviewed about race (all persons of color) remarked that race was the salient factor in understanding how they had experienced life on campus and within the research environment. It is almost as if racial sensibility was more clearly inflected through a cultural and historical understanding, while gender was more *naturalized* and *operated within a system of trade-offs* (Keller, 2001). Although "*flirting*" or gaining favors from a paternalistic mentor may both be fraught with pitfalls within science practices, they can also benefit a woman. By contrast, one cannot "flirt" with race, nor did any person of color speak of being smothered with White mentoring attention. However, both groups spoke of meeting up with White male condescension. In the case of gender, being a "girl" is familial and may also connote being "dumb," whereas condescension from majority faculty toward persons of color or student peers was usually experienced as exclusionary, not familial.

Therefore experiences in the lab related to gender, in comparison to those related to race, were more elusive, glimpsed, yet also somewhat disclaimed. However, we should note that our interviews allowed participants to bring up gender issues on their own. It can be uncomfortable for women to identify with feminism (Kerr, 2001), even if women scientists with more experience typically can cite various types and instances of gender discrimination in their own lives (Sonnert & Holton, 1995). Gender questions in our interviews were open ended and emerged in relation to other lab issues rather than through a directly focused, structured interview format. In contrast to race, especially for African American participants, gender may not be at the forefront of conscious, or at least reported experience; gender was treated more as an effect of natural, taken-for-granted differences. Even so, the women interviewed spoke of their experience in ways that suggested that they position themselves as operating somewhat differently from men, both socially and instrumentally.

Although interview questions pertaining to race were open ended, they were derived from focus groups that had discussed racial issues in school and science. Race seemed quite salient to our participants. This might reflect the fact that, in comparison to gender, there is a much clearer history of the meaning of race in Western culture, even if one cannot always be sure that race is at issue in a given interaction. In general, participants in the study on race seemed to have greater knowledge of racism and its possible effects than knowledge of possible effects of gender.

Consequently, Part I on *The Person Enacting Race* and Part II on *The Person Enacting Gender* deal with the different ways our participants experienced what they considered to be quite different dimensions of identity. It may well be that there is overlap in how any person deals with marginalization or with what it means to be positioned as the "other" in another's eyes.[1] However, in this text, we prefer to treat the issues of race and of gender as distinct conceptually and functionally, and differences in our data support this distinction.

PART I: THE PERSON ENACTING RACE[2]

The psychological study of race and racial categorization is a relatively recent endeavor for social psychology. In its earlier history, psychology's relationship to race was deplorable, the history of intelligence testing being a ready example (Guthrie, 1998). Recent efforts by psychologists to address race often leave explicit study of this history behind, even if meaningful interpretation of research data requires it as background (Eberhardt, 2005). Some authors have explored the resilience of African Americans in the face

---

[1] It is important to clarify that "other" in this context refers not necessarily to the actual composition of the laboratory in question. Rather, it refers to how one is positioned with respect to the *culture* of science as it has developed historically; this is a culture that endures even as women and persons of color increasingly enter the laboratory.

[2] The methodology and sample for this section are somewhat different from the sample drawn on in the other chapters. In addition a number of passages from participants are from the five focus groups on race and science for which we were not always able to distinguish the exact identity of participants (just that she was different from others in the group). Thus we use the convention of participant 1, 2, 3, etc. See Chapter 2 on methodology for further specifics on the sample and method.

of enormous oppression (e.g., Jenkins, 1995); others have examined the contradictions of African American life as generative of new possibilities (Neal, 2005); yet others have begun to explore the interdependence of White and Black identities by seriously interrogating whiteness (Fine, Weis, Pruitt, & Burns, 2004). In addition, there has been a lively and informed discussion of racial identity in a popular/professional set of publications that draw on psychological research but are aimed at the wider community (Jones & Shorter-Gooden, 2003; Tatum, 1997).

The disciplinary mainstream, by and large, examines prejudicial attitudes and their enactment in the context of experimental social psychology and a framework of social cognition using the concepts of stereotype threat, implicit association tests, and fundamental attribution error. Although there are exceptions (Young-Bruehl, 1996; Horkheimer and Adorno's critical theory analysis, 1947/1973), psychology typically treats prejudice as an aggregate of attitudes "inside" the person, evoked by certain conditions. However, the focus on prejudice as a possessed attitude – instead of, for example, how one "performs" whiteness – is challenged by social identity theory's conception of identity (including racial identity) as a phenomenon of intergroup relations. Therefore social identity theory understands marginalization of any group not as a matter of aberrant personalities or deficient cognitive processes, but as intrinsic to group-making itself (Dixon, 2007; Tajfel & Turner, 1979)

Similarly, an important assumption in this text is that a psychology of race should illustrate its fluidity, which is required when race is construed as a social enactment rather than an attribute. In bringing race to the psychology of science, the task becomes one of analyzing where and how race is enacted in scientific communities such as laboratories. Given that science and rationality are presumed to transcend social and historical factors, the question becomes particularly salient yet precarious. Perhaps even more powerfully, scientists are historically juxtaposed to less rational categories of being, women and persons of color among them. Thus tracing how race is reinscribed through positioning within a given laboratory and how a minority member might position him- or herself in response is an important contribution to a broader understanding of what it means to be a person of color. Race becomes a way of reading the effects (rights and duties) of being positioned in a particular manner: implicitly, through associations and representations, or in terms of what we refer to as the broader social pact of the laboratory community.

As we have seen in the preceding chapter, the effects of positioning are epistemic as well as social, and this is particularly relevant to the study of race and science. In this section we focus on the ways in which scientists of color, in this case science students, are racialized through patterns of interactions that seem to keep them on the outside of the laboratory community's formal and informal networks. This dimension of laboratory life is then understood in terms of the self-positioning of the students (away from research) toward industry and toward their racial identity and in relation to their motivation for working within the lab.

However, the analysis of rights and duties in conversational exchange that accompanies positioning theory becomes more complicated and layered when dealing with participants who are marginalized. A participant's sense of legitimacy as "one who can speak and know" is compromised more easily; one can encounter a discursive erasure that is fundamental. In discussing how positioning theory needs to address the question of excluded groups in specific ways, Wilkinson and Kitzinger (2003) asserted,

> One of the challenges facing positioning theory (and conversation analysis) is to analyze more systematically the ways in which the endurance of a taken for granted social world ... [or] normativity is produced through certain encounters in which the positioning of lesbians, the disabled, other ethnicities, and so on is routinely effected without anyone setting out to do anything particularly "prejudiced" and in pursuit of another agenda entirely (p. 172).

Kitzinger and Wilkinson, well known for their work in conversation analysis and in the psychology of gender and sexual orientation, point to an implicit creation of normativity – a sort of social pact that undergirds community – which determines the range of positions one *can* take and which is so taken for granted that it escapes comment. In this chapter, we see how being positioned as outsider and seeing oneself as an outsider implicate race in science practice. In particular, we examine some moments in which minority participants appear to feel prohibited from speaking about race, but can draw few explanations for certain experiences of exclusion and invisibility other than (non)conscious racism by majority peers and faculty. When one's sense of oneself as a scientist or science student appears irrevocably inflected through racist or at least racially grounded responses, this perception can reinforce identity as a person of color as an alternative position that supports one's goals and activities.

Thus we pursue somewhat trickier material relating to identity here (i.e., not when identity facilitates sense-making, as is emphasized in the previous chapter, or when identity effects lead toward coherence; see Redman, 2005).

Rather we examine positioning and identity in terms of impasses and conflicts, which are analyzed in ways that refer to the question of how one is perceived within the social pact that creates the possibilities for positioning and supplies the social affordances. Sometimes this means that a student will discuss not feeling able to gain entrance into the brokerage of rights and duties, no matter how locally circumscribed.

Discourses produce rights and duties, including a bid for warrant in speaking about the world (science discourse). Yet without affirmation from the other, identities degenerate.[3] Each conversational exchange in the laboratory involves a discernible system of meanings, representations, and ideologies; the exchange affects the net of meanings that situate participants. Yet there are frequently conflicting systems affecting and being affected by speech, as well as other dimensions within speech that are alluded to and aimed at in speaking but that remain in a way unspeakable. Some conflicts, if not unspeakable, cannot be spoken well.

We begin this discussion by situating diversity as a problem relevant to science practice and then turn to a more detailed analysis of the discursive practices of our participants.

THE PROBLEM OF RACIAL DIVERSITY IN SCIENCE

An abundantly documented history confirms that racism and sexism in science have affected practices of conducting research and selecting topics for investigation.[4] Both women and persons of color have often been cast as the curious objects of science rather than its practitioners. Many in fact see the very category of race itself as traceable to a certain moment in the history of science (Shepardson, 1998), in that science played a pivotal role in creating racist and sexist categories to organize differences between peoples and justify subjugation.[5] Unfortunately, scientists have not evidenced a particularly astute understanding of this past (Rosser, 2004; Schiebinger,

---

[3] The French author Franz Fanon, whose work spawned an entire approach to race in cultural studies and within the social sciences, refers to the look by Whites that disrupts the visual field and reflexively abrades the body. Reality as a lived experience, as action in the world, is punctured by this look. Fanon refers to this manifestation of racism as a corporeal malediction, wherein one's corporeality is *not* given by "residual sensations and perceptions primarily of a tactile, vestibular, kinesthetic, and visual character, but by the other, the white man, who had woven me out of a thousand details, anecdotes, stories" (Fanon, 1967/1952, p. 111). See also Gergen (2001) and Harré, Moghaddam, Cairnie, Rothbart, and Sabat (2009).

[4] See Guthrie (1998); Sinclair (2004); and Harding (1996).

[5] See Weigman (1998); Laqueur (1990); and Richards (1996).

2002). Instead, there has been an effort to rectify past inequities using various approaches for enlisting underrepresented minorities into the scientific fold. Yet even after decades of proclaiming a desire for diversity within science, researched and encouraged by organizations such as the National Science Foundation in the United States (Congressional Commission on the Advancement of Women and Minorities in Science, Engineering, and Technology Development, 2000), advancement has not been achieved to the extent required. Full racial diversity at research universities, at least, has yet to be achieved. For example, from 2000 to 2007 the percentages of White graduate students in science and engineering only dropped 5 percentage points (from 71% to 66%), and very modest gains were made by minorities at all levels of engineering graduate schools (Burns, Einaudi, & Green, 2009; Leggon, 1997).

Donna Nelson's (Nelson & Rogers, 2002) data on minorities and women at the faculty level, which are in the form of raw numbers, show how inadequate current attempts have been in retaining minorities in academia and university research. Minority women tend to leave academe for industry and are disproportionately represented in institutions with higher teaching loads (e.g., community colleges, teaching colleges, comprehensive universities, and historically Black colleges and universities).[6] As a result, Nelson and Rogers noted, "[I]f the student is a woman of color, it is probable that she will earn her Ph.D. without ever seeing a minority female professor in her field (Nelson & Rogers, 2002, p. 2). Put bluntly, there has been a discrepancy between the goals of diversity articulated by majority research universities and the ways in which such goals are "instantiated within the work setting" (Olsen, 1995, p. 287).

Many earlier, "let's fix the problem" efforts targeted recruitment in what was referred to as the "pipeline" approach. However, current opinion about the merit of these efforts is skeptical. "[T]he pipeline empties into territory women and faculty of color often experience as uninviting, unaccommodating, and unappealing" (Trower, & Chait, 2002, p. 34). As the pipeline approach has come under increasing scrutiny (Alper, 1993; Xie & Shauman, 2003), several culprits have been identified as the causes of failure to retain minorities. These are arenas in which social psychological, and particularly discursive psychology frameworks, have an important role to play.[7]

Seymour and Hewitt (1997) conducted an extensive study on the experiences of science, technology, engineering, or mathematics (STEM) majors, particularly on minorities' and women's decisions to remain in a science

---

[6] See Moody (2003) and Hamilton (2004); and Mannix (2002).
[7] See Eberhardt (2005); Devos & Banaji (2003); Steele (2003); and Gee (1999).

major or switch majors. They looked at classroom climate, mentoring, faculty, and peer relations and the effects of a variety of outside social factors in determining a given student's decision regarding his or her major. The allusion to "outside" social factors (Nasir & Saxe, 2003) is essential to the following section on gender as well. So here as we look at the contemporaneous integration of identities (or trace identity effects; see Korobov, 2010) within science for minorities, we find a somewhat more combustible mix, a more complicated configuration of positioning. Yet the difficulty should not dissuade sustained inquiry. Recent research with minorities at the undergraduate level has suggested that "identity"-related categories correlated to "seeing oneself as a scientist" and "being recognized" by others as such (one can see where positioning is at stake here) are important factors in successful mentoring experiences of underrepresented minorities (Brown, 2004; Hunter, Laursen & Seymour, 2006). Thus in both explicit and implicit ways, the ways in which one positions oneself and is positioned by any given community are essential to the broader psychoeducational issues at stake in minority retention in science and engineering.

Our work in this section stems from such concerns and falls into the broader set of efforts to articulate the informal events, encounters, and interactions that work against the goals of increasing diversity in science (see also Hunter et al., 2006). As discussed in Chapter 2, the following data come primarily from interviews and focus groups in which the majority of participants were African American women at various stages of their academic careers, from master's student to faculty member.[8] In keeping with other significant research in this area and with the aims of this text, we geared our interviews and discussions to explore real-time experiences of underrepresented minority women in science (Carlone & Johnson, 2007; Johnson-Bailey, 2004; Olson, 2002). The unit is the acting person as she sees herself situated *within the lab* and as that person is emergent and constituted within interactions or practices. Here we cull the identity effects of such interactions in relation to the lab's practices, projects, and interactions. As the previous chapter shows, such identity effects are ongoing and intrinsic to learning and cognition.

RACE AND SCIENCE

Bridging social and cognitive systems and processes through the analysis of the acting person in normative contexts of practice is a goal addressed throughout our text. For the most part, in previous chapters, we have

---

[8] See also Malone and Barabino (2009) and A. Johnson (2007).

brought forward many seemingly nonscientific factors (e.g., emotions or identity) into the analysis of practices that are integrated within the internal workings of the lab community and its problem-solving goals. The focus in this chapter remains the same, but the range of representations and social factors widens to encompass the broader socio-cultural context of race relations. We examine ways in which dimensions of identity usually seen as external or irrelevant to science practices are re-created within science practices and contribute to any person's identification as scientist (as it is contingently enacted), her motivation, her cognitive practices, and her perceived place within a given science community.

Among those participants who spoke about racial issues, there was a keen and unequivocal awareness of being positioned and positioning themselves in relation to majority faculty and students both inside and outside the lab. Thus one student talked about eating egg rolls with her traditional turkey on Thanksgiving as an overture to becoming more involved with a lab where there were mostly Asian students. Another spoke of the "delicacy" of getting drunk while around Whites and the stereotypes doing so might evoke. Yet another nursed a beer all night at a lab hangout and described it as building better relations with her lab mates. All noted that informal gatherings can lead to important exchanges of information: Many tricks of the trade, such as which materials and procedures work well, which articles to read, whose boyfriend writes computer programs, or who are the knowledgeable people in other labs, are examples of information supplied through such informal conduits.

In addition to awareness and discussion of how they were positioned (if not in these terms), the participants explicitly considered how they could better position themselves. Sometimes this meant forming alliances with White lab mates, but doing so is complicated by feeling excluded from networks of those types. Further, it was recognized that *denied social meanings* were enmeshed with the rituals and practices that sustain the community of learners. Often such denied social meanings impact one's progress in a science career, the speed of one's problem-solving and presentation of results, and even motivation for scientific practice. In the following, a postdoc reflects on her experiences of gaining a position at a lab[9]:

sp14: *Yeah and I have that, I had that with interviewing, I interview with an old White male guy, and I'll just be like, I don't really know what I'm*

---

[9] The labs in our ethnographic study did not represent a sufficient population for this analysis. There was only one African American researcher (female, Lab A).

going to say, [laughs] *hopefully, they are nice, and I'll be able to sustain the conversation, but . . . it was definitely scary.*

An important distinction to introduce at this point is what Kinder and Sears referred to as that between explicit or intentional prejudicial attitudes and *symbolic racism* (1981, p. 416, emphasis original). and intentional prejudicial attitudes. Kinder and Sears analyzed the effects of symbolic racism on political attitudes, but later researchers have extended the concept to other contexts in which implicit racism may be revealed, including the academic workplace. For example, Olson notes that symbolic racism denies "overt forms of prejudice while denying access to resources, information, and sources of support" (Olson, 2002, p. 271).Thus we might see symbolic racism in the context of minorities recounting feeling and being left "out of the loop" (see also Clark & Corcoran, 1986). No overt bigotry is expressed; one is simply forgotten or omitted.[10] In the words of one of our participants:

sp4: [How do I know if] *[t]his person is not gonna screw me and this person has my best interest. You know, sometimes it's not even – and I've learned this most recently – it's not even a deliberate trying to screw you. Sometimes it's unconscious and they don't know that, that they're not as forthcoming with information as they are with other people. And it's not like they're – you just didn't cross their mind. Or they never thought to share that with you.*

The idea of symbolic racism can be extended to include the particular ways in which social meanings and interactions are coded and understood in the wider culture and are insinuated into the everyday interactions of STEM faculty and staff. These broader meanings, underlying various acts of enunciation and social exchanges (which involve when and how something is said as well as literally what is said), implicate the place of women of color within science more generally. Regarding the place of minority women in another educational environment, Juanita Johnson-Bailey noted,

When we participate in programs or classes as students, instructors, or planners, we bring the historical weight of race with us. It matters little whether we intentionally trade on or naively discard the privileges,

---

[10] In *The Racial Contract*, philosopher Charles Mills brought this question of "speaking rights" into concrete expression. He noted James Baldwin's remarks that "if I should speak, no one would believe me," and paradoxically "they would not believe me precisely because they know what I said was true" (Mills, 1997, p. 97). The status of one's spoken word complicates the positioning strategies of minority students, who discussed in one of their focus groups how speaking up about racism left them – in their labs – being seen as Black only (yet again) and wanting "special treatment."

the deficits, or standpoints of racial statuses. Such ranks, authorizations, honors, suspicions, and stereotypes cannot be set aside. They are accrued in society's invisible hierarchal banking systems of trading and bartering according to designated racial rankings. If teachers are to function proficiently, they must acknowledge and manage the uninvited specters of race that haunt our practices (Johnson-Bailey, 2002, p. 40).

As Johnson-Bailey has suggested, these historical factors do not simply disappear when they are supposed to, but instead speak through the person in the evocation of given signifiers and *principally through how one is positioned through interactions.* This positioning may be partly embedded in practices such as mentoring and knowledge transmission or in other informal practices through which one is inducted into the science community. Thus the question, "does one belong here" (in science), is not an abstract question for scientists or science students, particularly for minorities, but is lived in everyday interactions (e.g., entering the lab and seeing you are the only Black person, being in the loop, hearing about conferences, being copied by a principal investigator on an email). Further, one's sense of belonging sustains one in relation to a research agenda: It recognizes, legitimates, and gives symbolic form to one's efforts.

More broadly, in science practice one's relation to others' intentions and desires are intermixed with one's relationship to knowledge, authority, and recognition (from peers and faculty). The following monologue, taken from a focus group, reveals shifts in a sense of boundaries and of feeling "outside" the lab. Although focus group participants spoke of not wanting to see things in terms of race or racism (which, as recent history attests, predictably sets off a series of responses from White persons), particular interactions would lead to an identification with their ascribed racial group. According to Alcoff (2006), black identity requires "not only that one is treated as a black person, or that one is objectively black, but that one is 'subjectively black' as well in the sense of actively interpreting the implications of this imposed category for one's sense of self and community life" (p. 244).

In this interaction, the participant recounts her exclusion from the social pact of scientific work and its very tangible instrumental effects on her experimental success and academic progress. As well, one can see her eventual positioning in alignment with her minority community. While speaking of navigating problems in her research efforts, she recounts her relationship to peers, faculty, and to knowledge:

PARTICIPANT 1 IN FOCUS GROUP: *I think we have this Lone Ranger syndrome, unfortunately. We come from a situation where everywhere else, we*

all have each other's back, we get in here, we get into this academic, we figure
we have to do it all by ourselves. You know, everywhere else we never do any-
thing by ourselves, we have all our friends and all our support people, but we
get in academia and we just figure, we gotta know all the answers, we gotta
be, you know, don't speak until you're absolutely sure. And that's one thing
I, I try to tell [a faculty member] *like going to conferences, you know, I meet*
someone and I'm like that's bull, that's bull, what she said, but this is what's
wrong and the, to kinda, not berate someone down, but just to build them
up like okay they are the expert but there are some holes here, your research
papers, but this is what they're leaving out, and it's not perfect, so even when
you try to repeat this experiment, don't feel bad that it didn't work, cause
you're gonna read hundreds of papers where. . . . they make it look like the
whole field is like that, but I mean you need someone to tell you that. They
don't tell you that in class, I didn't have a mentor in lab to tell me this, I
need someone else to say that. . . . do your research, your research is what's
important, you know, but I think that's the kind of mentoring that I think
other students are getting from each other that we're not getting because
we're not in their inner circle or whatever and we don't feel comfortable
cause we figure we have to know everything, cause they know everything,
but they don't. They're all sharing information and we're not in the sharing
pool so. . . . [focus group talkover] *. . . we need to be mentors to each other.*

It is very important to emphasize that the relationship between these "social" qua mentoring qua community relationships and one's cognitive work is intrinsic, not an add-on required by a special category of student. In mentoring processes, important information is passed along, knowledge is produced more quickly, and one becomes authorized to work, fail, and succeed.

This cognitive dimension of the mentoring relationship can be clarified through Michael Tomasello's research in language acquisition, which locates the cognitive properties of language in the intersubjective realm he has called "joint attention" (2003; see also Tomasello, 1999). The interactive space of joint attention, including what one appeals to in the other through speech, is important both for accruing membership in a community and for feeling authorized as an independent thinker (e.g., as belonging to the set of scientists). For Tomasello, it is important for thinking more rapidly and productively: Effective joint attention accelerates the production of knowledge. In science practice, there is a specialized language with rules and idioms that are ostensibly transparent and abstract (i.e., strictly about science).

Nonetheless, as we see repeatedly, in the real time of science, social representations enter the mix. As Butler (1993) suggested, the social, personal, and cognitive dimensions "materialize" within a given practice: a moment of insight or communication of experimental results. A senior researcher's ability to say to a student, "Yes I can see it that way," lends authority and legitimatization to that student's perceptions and acts, which in turn complement the discourses that render rights, duties, and obligations within a laboratory. As our participants note with great emotion, such authorization of their work requires a series of social maneuvers in relation to authority, in addition to the clear presentation of data or results. Standing up to the PI is a learning experience, and gaining an independent sense of one's research trajectory is discussed with great vigor by the participants. This issue is not confined to underrepresented minorities, but is treated differently in the context of racial and ethnic differences.

POSITIONING, MARGINALIA, AND THE SOCIAL PACT

Consistent with the framework of the acting person in normatively structured frameworks, the acting person of science always participates with and operates in different layers of an implicit social pact. There is one sense in which the social practice is a local one, peculiar to the social relationships and hierarchies of the laboratory in question. Yet there is another dimension of the social pact within any university laboratory that is rooted in the history of science, to its normative operation.

Every science graduate student has to negotiate entrance into the local science community of the laboratory and the wider science community within which each lab is situated. For less well-assimilated members, there are inherent conflicts, exclusions, and impasses in this process.

Thus, racialized persons entering a science lab can encounter tensions related to "being in the loop" as a mode of access to information and resources. These tensions are seen in how they find themselves positioned in relation to the legacy of science and in their own desire (goals, aspirations, interests) in relation to this legacy. This desire is implicated and satisfied when one's discussion of one's research is recognized (thus validated) by another or when one's projects and interests are coordinated with those of the broader community. If one feels positioned as outside the realm of legitimacy in one's practices (and desires), one will turn to other audiences for recognition, to other interactions in which one's perspective or self-image is recognized or reflected back. The levels of disenfranchisement are longstanding and multilayered, with neither majority nor minority students controlling the currency. However, it is true that responsibility falls on those with the most cultural capital to effect change and open up new affordances for minority students in daily research interactions.

The following excerpts are from an interview with a doctoral student who was starting her dissertation research, but who encountered a number of difficulties integrating with her lab and lab mates. She had talked earlier of going into industry rather than pursuing an academic career and research. As noted in the literature on science education, many underrepresented minorities and women do move into industry.

I: Do you feel some pressure to, every once in a while I feel like you've got this voice that's saying, stay, stay.

SP12: *Absolutely, absolutely, you're like, that's what I was just thinking, It's like you're here, this is a good opportunity, you should, you know, try to stick it out, you know, this is your burden, you know, you can do it, you know.* [I: Yeah.]

SP12: *You're in the position to do it, so.* [I: Yeah.]

SP12: *You should do it, um, but then I see the like I was about to say the few minority professors that are here with doctorates that are working, to me, they seemed secluded [by] themselves. And, and I think. You know I just couldn't imagine volunteering for that anymore, you know.*

The same participant then speaks of the effects on her research of her isolation, not being in the loop, and lack of recognition. She has had to push her PI to have meetings with her, although he holds regular meetings with his peers. She questions whether her research is of interest to him and seeks faculty and peers outside of the lab to help her with the inevitable glitches that come up in experimentation. She continues,

There's, there are certain things that you just you have to know and there, they're not necessarily in the books, they're like out of techniques that that people who've been doing, who were biology majors in undergrad, you know, they know that. You get this kit and you're gonna automatically off the rip have to tweak it a little bit to make it work.

The absence of this everyday help and collaboration positions her as an outsider to the science community and takes its toll on her research. She presents a more precise account of her isolation in its effects on research:

SP12: *Yeah . . . so, I don't know. I think, a big problem too and I – we've talked amongst ourselves, many of the people in the focus group were really good friends, when you have these experiences it takes away the passion for your research and I think that the true researchers that stick around are people that, the professors, they have a sincere passion for research and you start off with the passion for it but when you take on all these other negative things you . . .* [I: Uh huh.]

SP12: *Kind of project it on to the research. And you're just like, and you get frustrated with it and then you don't, you know you wanna do it, it wanna do well.* [I: Right.]

SP12: *At it, but it kinda taints the whole research atmosphere and you start associating all those experiences with doing you know this type of work and you just kind of. I mean I really am interested in my research, but I think it's definitely affected my passion for it.*

Another participant articulates this cognitive/social cusp in different but equally telling terms:

SP10: *That um, sense of being apart from the group. And so, wh – what that does is, it affects you in a certain way that you, you don't open up as much as, because, granted, you always have a classroom and you always have a formal way of learning, but you also learn informally, so and it, it is um, supplemented, if to say and so like, so you can learn* [inaudible], *but then how many people around you in the lab that you can talk to them about your research, and if you're feeling separate* [inaudible], *will you be able to open up to your weakness and, and have them – not criticize it, but try to strengthen it. So, now if you feel that you're different or you feel you're being treated differently, you become insecure about those differences and don't want to expose them, then not, not admitting that you're ignorant of something, you'd never seek out . . .* [I: Yes.]

SP10: *. . . the knowledge, and you never know, you wouldn't know more.* [I: Yeah.]

SP10: *And so I think that affects me a lot in my research, because then I – I withdrew?*

One's generalized capacities for belonging to a community are often expressed in very subtle ways. The following narrative recounting of a set of exchanges between a minority participant and her majority peers in her research lab is illustrative:

SP3: *Yeah, I just, you know, don't, I don't talk you know, is just like okay well, I actually didn't mean it that way, whatever, you know, I don't care to explain this situation to you. You know, I mean, even when you know, everyone's in the lab trying to joke, you know, if I crack a joke like someone will freakin' respond to it like I'm some kind of idiot, you know, okay, come on, you know, I've got news for those people, I'm not a complete idiot, but you know, it happens like that, you know. I remember one incident, it had nothing to do with academics, I think we were talking about the W. hotel or something, and I was like, Yeah, I saw this bumper sticker. I was like, I don't think I could stay at the W. hotel, because I saw a bumper sticker that said uh, "W" the President, and I was concerned that there was a relationship and they were like he doesn't run the W. hotel. I'm like, I know that, I mean, come on!*

Another participant speaks of not sharing "common ground" with her advisor, saying, *"It's unlikely that my advisor's gonna come and joke to me."* The number of levels at which humor and similar interactions function subjectively is important. A joke requires there to be a shared relationship between interlocutors to draw on the resources of language. Because a joke requires considerable linguistic resources, not being qualified to make or take a joke is a sort of denial of a certain symbolic latitude, which operates at a fundamental level, like being addressed by your proper name.

Another fundamental issue, as referenced by the student quoted earlier, is that of being recognized by name:

> SP10: *And (laughs) you know it's funny because I go to um, this conference, the X Research Society – X Research Society conference, and I go every morning, every year for maybe the last five years, and my advisor's advisor usually has a party . . . for all his former students and their students, so just to get the connection going? And believe me, I'm the only Black person there, but they never remember me every year. I'm like that is really funny. . . . You know? Because I think I would stand out to them, you know?* [laughs]

This anecdote is representative of other participants' recalling a failure of majority faculty and staff to recognize them by name or confusing one Black person with another. One minority faculty noted that both he and another minority faculty are married to majority women, and colleagues at his institution had called his wife by the name of the other faculty member's wife. This mistake communicated a certain racial vigilance on the part of majority colleagues, who apparently linked these women by the race of their spouses, rather than by their proper names or resemblance. Minority students often report being confused for one another or simply, after numerous introductions, having their names forgotten.[11]

---

[11] The social pact can be understood as containing what Mulkay and Gilbert (1986) described as asymmetrical discourses: science proper and science improper. The latter is full of the politics, unsupported yet essential intuitions and insights, character flaws, and other factors that both advance and impede what we can say about the world. Attempts to control the effects of what is not proper are revealed through an official discourse and methodology. Yet the proper cannot exist without the improper. The discourses of proper and improper science support one another in the intuitions and alliances that create knowledge; therein we might speak of the cross-fertilization of social and cognitive dimensions. But equally, the improper serves to maintain the *boundary* of the proper. The dual relation structurally affects how minorities attempt to position themselves and how they feel positioned by majority institutions, leaving them to negotiate their positions within a set of parameters about which their White colleagues and mentors do not or cannot grant belief.

There is a fine line between not having one's name remembered and not being recognized or being visible to others in a broader sense. The following interview passage suggests an experience of invisibility:

> SP15: *Because there was this naïve opinion of, if you can just get them to come, they'll succeed and they'll stay and they'll graduate. Now, you get them to come, they'll feel isolated, and nobody wants to be their friend, because they didn't go to the right school, they didn't pledge the right fraternity, they don't look the right way, and everybody else just pairs off with everybody else, and you're just sitting there by yourself. It's amazing when you're sitting in a room and they say, trade places – papers with the person next to you, this person beside you trades with that person and that person trades with that person, and you're still left with your paper.* [laughs] *And you're like* [I: Yeah.]
> SP15: *Why did you look that way, why did you look . . . do I smell?* [laughs]

The manifold ways that minority students may feel left out of the loop or not taken seriously would require a longer exposition, but such perceptions, which happen in contingent and routine interactions, are different from responses to explicit or overt moments of racism or exclusion. Yet they hit at the very heart of the social pact. Thus, one need not focus at all on explicit racism, which still exists and easily arouses the attention of minority and majority students and faculty. Nor do we presume that the history of racism in science is imported wholesale into everyday practices, though to borrow from Judith Butler, historical significance frequently can be "cited" in present-day interaction (Butler, 1993). Rather we assume that symbolic disenfranchisement (bearing some relation to the history of racism in science) affects one's symbolic capital and the intricacies of positioning in the laboratory. These subtle forms of disenfranchisement can be easily experienced as racism.[12]

As with the joke and calling someone by his or her proper name, certain aspects of the contract of speaking, even casual ones, are key in bringing members of the community into the fold of learning and knowledge production, as seen in the following:

> SP10: *But then it was because I had gained more confidence by the time I was able to do that, because I know that, okay, it hurt me a bit that he* [the PI] *peeked into the room and talked to everyone else and I kinda set it aside, and*

---

[12] The material in this section draws from narrative data gathered during a study of gender and science in research laboratories that was conducted during the same time that we were working on cognition and learning in two major biomedical laboratories. The sample is drawn from the same population as the cognition and learning study. The methodological approach is described in Chapter 2.

he didn't come to me, so I told him you know like, You, why don't you say hi, you're passing by, and you have time to tell everyone else?

It is not trivial to a person of color to be the only person not spoken to in a lab or to be the only one whose contribution to a lab meeting is passed over. Research in gender has demonstrated how power differentials (including one's social status as a woman) affect whether one is acknowledged in speaking (Valian, 1999). There is much less research on minorities in this regard. Yet our participants were aware of experiences of not being "heard." As one participant (S10) put it, "*If I don't make an impression at all....*"

In response to queries about invisibility, persons of color in our study remarked that they often felt that they were not heard or included in informal and formal networks. This sense of symbolic exclusion is noted in the literature (Jordan, 2006; Myers, 2002) and it communicates a sense of feeling invisible when it comes to their contributions as scientists. It is a way of being positioned on the outer circle (Zuckerman, Cole, & Bruer, 1991). For example, their contributions to lab meetings are sometimes passed over, but when the same idea is expressed by a White peer, the contribution is acknowledged:

SP12: *It's like can we rewind and I'll put it on megaphone and say did you hear what I just said, oh absolutely, absolutely, and you know I often, uh, project that on myself, like to say well did I not speak clearly enough, did I not say it loudly enough, did I not command the attention, like what am I doing that's not...* [I: Right.]

SP12: *Right you know* [I: Yeah, yeah.]

SP12: *Getting the point across, because like surely they are not intentionally doing this, you know, I, I honestly think sometimes that um it just gets filtered, you know, you see who's talking and you kind of, people will pay a little less attention, so.* [I: Yeah, yeah.]

SP12: *It just kinda gets filtered and then someone else says it, it doesn't have to be a mind shattering thought or statement, it just...*

I: You mean a matter of...?

SP12: *The filter becomes, turned back on because this person is not speaking, so, I think it just kinda, kinda works like that, I'm not exactly sure why, but... yep.*

I: Do you think it has something to do sometimes with race or do you think it's...?

SP12: *No, I mean, yeah, I'm saying I'm not exactly sure why we're still dealing with that, oh but absolutely. Yeah, I'm just thinking at this point, you know, why would that still?*

I: You're thinking, yeah.

SP12: *Be the case, but, oh absolutely, absolutely, race more so than gender.*

The reflections expressed by SP12 are echoed by another doctoral student who is finishing up her work in the lab. Other participants in the focus group readily agree with her remarks:

> *When I first got in my lab, I had that a lot* [invisibility] *and I look, like didn't I just . . . [Didn't you] hear me say that?? I just said that* [respondent sounds incredulous]. *You changed just one noun, maybe one adjective and everyone jumps on it.*

Exclusion from informal and formal networks complicates one's relation to the lab and lab work. It not only forecloses many avenues of the transmission of information that would foster agency and learning, but it also hampers a sense of legitimacy, which is equally needed to navigate the complex challenges facing any graduate student. The encouragement of supportive informal networks by PIs or lab leaders is critical, and whether students of color feel less related to their advisor or PI and whether they experience isolation can be important *epistemically*.

It is important to note that mentoring relationships should encourage the ability to appropriately challenge lab authorities and to stake one's claim to knowledge. Here students discuss the positive outcomes of confronting the PIs:

> PARTICIPANT 1: *Because I was doing that, I was talking to X about it, I was like, Well, this is what he wants me to do, this is kind of what I want to do, I'm a little afraid to you know, just say I'm doing this, because then I might not get your support. And you know, we kept going back and forth. And I finally said you know, Dr. W., this is what I'm more comfortable with doing, this is what I'm more interested in doing. And all of a sudden, in the next meeting, he was like, Ok, well. . . . and then we just started moving,* [Others: Uh huh.] *and I was not expecting him to do that at all.*

> PARTICIPANT 2: *It's a test! Maybe it's just a test, a rite of passage. It's the push to see when you're going to push back. When you look at them as a colleague, or look at them you know, instead of Almighty, and say,* [4: Yeah.] *you know what, I DO know something.*

SUMMARY

The experience of our participants indicates that race has subtle implications for the everyday practice of science and science education. Most notably, issues of authority, recognition, and inclusion are interwoven with the ability to engage successfully in the cognitive and social practices of a community. Attention to explicit and implicit inclusion as expressed in interactions with other researchers and in the attitudes of PIs is crucial within this difficult terrain.

PART II: THE PERSON ENACTING GENDER[13]

Questions about gender and science traditionally are framed as questions about bias; namely, whether the structures and practices of science are systematically less hospitable to women.[14] A less examined question is how gender might be enacted in the day-to-day practices of science; that is, how are gender differences in science lived out?

As noted in the introduction to this chapter, whether or not one can easily discern gender differences in science practice is a contentious question (Longino, 2001). There are documented gender differences in some realms: Women are less inclined to enjoy competitive boasting, respond better to less cutthroat teaching environments (Etzkowitz, Kemelgor, & Uzzi, 2000; Seymour & Hewitt, 1997), and tend to be more attentive to details (Malone, Nersessian, & Newstetter, 2005; Sonnert & Holton, 1995). Fox and Mohapatra (2007) suggested that, although productivity relates to gender in some contexts, one must take into account the productivity and gender composition of the entire lab and not just the gender of a given scientist. Emphasis on gender composition begins to suggest that gender should be regarded less as an individual attribute and more as an interactional variable. Gender has to be considered in its wider situation of enactment.

Yet the very question about differences in productivity obviously stems from a perceived difference in what or how much is produced by men and women. Do differences in productivity simply mean that women do *less* science, or do women *do* science somewhat differently from men? Similarly, what is suggested by the documented evidence that women more frequently than men leave academics for industry careers? Does industry represent

---

[13] Of course opportunities for women are in principle expanding in every direction. Yet despite changes in the status of women in science overall, some subfields remain dominated by men (De Welde, Laursen & Thiry, 2007; Schiebinger, 1999). There is, moreover, a certain stubbornness in the process of integrating women into science. The aggregate numbers related to what is called "significant participation" still suggest that women do not move up the ranks as easily and thus are not often found in leadership roles in science (Burrelli, 2008).

[14] See Etzkowitz, Kemelgor, Neuschatz, Uzzi, and Alonzo (1994); Fox (2000); and Harding (1986).

a more family-friendly situation for women than academe (Eisenhart & Finkel, 2001), or is the less inviting environment of academics indicative of something about the practices of science beyond its rather unyielding family policies (Trower & Chait, 2002)? Are women somehow misfits in the normative contexts of academic science?

Questions about gender differences in the actual doing of science – scientific practices – concern the interrelation of cognitive and social processes, despite the tendency in science studies to separate questions about cognitive processes from social accounts of science (Longino, 2002). Although some scientists see their work as "gender blind" (Kerr, 2001; Traweek, 1988), others, as well as some sociologists of science and feminist philosophers of science, *do* see a possible difference in science practices, both in terms of the type of questions asked and the style of doing research (Haraway, 1991; Keller & Longino, 1996). For these scholars, nuanced differences in approaches to scientific questions may indeed reflect differences in gender socialization (see Rosser, 2001, for a summary). For example, Rosser's (2004) research indicated that women scientists do not always recognize climate difficulties in the workplace (which are discussed later), but that a sizable portion report feelings of isolation among colleagues. Members of our sample, as noted, often said that they did not notice many differences in method or intellectual practice or really much difference in any aspect of their work in the lab. They attributed most differences noted between women or between women and men to "personality." However, some participants clearly sensed gender differences even if they could not articulate their relation to actual laboratory practice. Moreover, "unclaimed" gender effects did emerge through accounts of their practices – as evidenced in positioning strategies, the importation of gendered metaphors in participants' narrative descriptions, occasional discrepancies in accounts given, and even odd slips and characterizations.

Just as the interrelation of gender and scientific work remains an open question within the literature, our own investigation of practices and experiences within biomedical engineering laboratories similarly provides no conclusive answer. An important point we wish to make is that, despite the broad-scale nature of gender as a sociological category, and any historical biases against women in science we might care to name, gender is a matter that appears to be personally framed and negotiated for each scientist. Our research yielded indications of gender differences worth examining and amplifying, but responses were heterogeneous. Admittedly our sample, though small, is large enough to enable us to say that we see no simple or straightforward relation between how one does science and how one is gendered. Yet, as elsewhere in this text, we find that the personal history, emotional experience, imagination, identity, and style of each scientist also enter the mix that enables scientific practice and thus affects gender enactments within it.

GENDER AS PERFORMATIVE

Thus the question becomes one of *how* the acting person negotiates the cultural category of gender within the normative frameworks of science practice, especially if we understand gender itself as performative, an enactment. Gender is a marker that a person manages in relation to cultural norms, power relations, and personal agency; it structures one's relations with others, at least in part. Gender is performative in two senses: It is enacted in relation to cultural expectations and ideals, and it has effects within a social situation that constrain and influence relationships with others (Butler, 1990). Given that gender is always being enacted in social contexts, it follows that at some points and in some ways gender and science practice intersect. Neither is static, of course. Each is lived dynamically through one's self-positioning in relation to a set of ideals and constraints as one locates one's creative ideas in relation to the rules and regulations of science procedure and negotiates one's particularity as a gendered person with the cultural history of gender (its representations). People do not "do science" or "do gender" in exactly the same ways.

Because the particularity of the scientist implicates embodiment, gender relates to the sex of the scientist, yet is not reducible to it (Connell, 2002; Scott, 1996). As articulated by Keller (2001), there has been a "double shift" in relation to theorizing about gender:

> This double shift in perception – first from sex to gender, and second, from the force of gender in shaping the development of men and women to its force in delineating the cultural maps of the social and natural worlds these adults inhabit – constitutes the hallmark of contemporary feminist theory (p. 133).

What are the implications of the shift? A shift in focus from sex to gender means a shift in focus away from merely counting the number of women in science. Sex is the presence of a certain sort of body, a biological given that is assumed to be expressed in social and psychological behavior (Fausto-Sterling, 2000). Adding female bodies to a laboratory environment may have salutatory effects on role modeling and the like, and there is evidence that this is the case. Yet it is not simply a matter of a percentage of bodies:

It is what the bodies *do* and *symbolize*. Putting more and more women into the pipeline to become scientists is helpful (Burrelli, 2008) but not fully adequate.[15] Here we run into what are essentially conceptual constraints concerning gender and science:

> [T]here is a tendency for gender, too, to become similarly boxed in or opaque. In particular, gender is often treated as a given category and one that maps neatly onto the sex of the particular individual or group being considered. Thus there is no conceptual space within which the construction of gender as an active process (for example through science and technology curricula and practices) can be explored (Henwood & Miller, 2001, p. 238).

The proper emphasis is not on a stationary attribution – sexed difference – but on what gender means and how it is enacted that might affect some women's ability to flourish in science. If, as Keller noted, gender is a map that organizes the world as well as designates bodies, actions are embedded in this mapping, occurring in how one *sees* (perception) as much as "what" one *is*.

Even with this more expanded gauge for understanding gender, it remains very unclear how scientific cognition is inflected through gender. Put differently, how are gender effects – a matter of social representations, personal history and motivation, and specific and characteristic modalities of action – implicated in and interrelated with scientific practices?

The perspective of searching out the underlying "causes" of women's slow integration into some fields of science situates our inquiry as directed to the "climate issue" that seems to haunt women in academe and science research (Conefrey, 1997; Ferreira, 2002). As understood in the field of feminism or education, "climate" refers to the social and intellectual environment in which women work (Ferreira, 2002). Clearly, tracking how gender differences are expressed through the person of the scientist, her place in the lab, and her research practices sheds light on climate questions and resultant issues about diversity in science.

Although the influences of the nuances of a given laboratory's social organization are not fully understood, scientists frequently note social factors as essential to their formation *as* scientists as well as to their (in some cases) disillusionment *with* science.[16] Equally important, labs, especially instructional labs, differ from other science and technology settings in ways

---

[15] See Ferreira (2002); Rosser (1999); and Seymour, Hunter, Laursen, and Deantoni (2004).

[16] See Lloyd (1993/1996); McIlwee and Robinson (1992); and Sonnert and Holton (1995).

that are significant for gender. Laboratory environments are more hands-on and often include cooperative learning such as apprenticeship, dimensions that may well be important in increasing diversity in science (Clewell & Campbell, 2002). There are often more mixed lines of authority among lab members. The ways in which peer interactions, differing educational levels, and varying levels of lab experience intermix to enable learning provide an interesting and unique environment for transmitting knowledge. The organizational culture is significantly different from that of a classroom or even instructional lab sections of science courses, where the "one who knows" is often more clearly demarcated. Further, in those settings students usually do not have to negotiate the use of materials and instruments – desks or notebooks – with other students; instructional labs try to avoid surprise problems and long periods of frustration and failure because of a lack of resources. All of the lab's differences from traditional learning contexts, especially the many points of informality in the laboratory, open up places in which personal *and* thus perhaps gendered ways of being are brought into actions and interactions. As we have tried to argue, gender is an inherent part of any activity system.

GENDER AND FAILURE

In a more general way, we might say that gender is part of how one enters the lab and how one positions oneself within its processes of knowledge transmission and creation. These processes include not only communication in the lab but also one's relationship to experimentation. Some feminist scholars argue that men and women have different working styles.[17] One aspect of working style might be how persons handle failure. Particularly within creative, innovative science settings, there are inevitable frustrations and discouraging setbacks at every turn, from failed experiments to poorly refereed papers. One also may not know why something works or does not work. Nature – and referees – can be unpredictable and unyielding.

In the cutting-edge labs we are investigating, members encounter numerous kinds of failure almost daily. Participants talk about 1- to 2-year lulls in figuring out what was wrong with what they were doing. Although there are some established procedures, researchers must often invent procedures and construct research artifacts as they go along. Experimental

---

[17] For the sake of clarity, we gloss some of its specialized Lacanian lexicon for the sake of accessibility, although some Lacanian nuances are lost.

technologies (i.e., the design of devices) are constantly evolving, as are the specific research goals and problems. Given all this, it is understandable that "*[i]t's 90% failure*," as one participant told us. Given the frequent failures and long periods of "intellectual purgatory," the question of motivation – a retention issue – becomes especially significant. As well, the complexities of acts are often better revealed at moments of failing. When we fail, we must rework our approach and must face, perhaps in a less articulate way, the personal reasons that motivate us to interrogate nature. When the existing system of knowledge cannot easily accommodate our research results, we must come to terms with our own particular web of investments; that is, we face the question, "Why am I doing this?" As we noted in the first chapter, normative science – and even pragmatism more generally – often ignores what does not work, what fails, and what is recalcitrant to adaptation and instrumentality.

In Mulkay and Gilbert's (1982) study of biochemists, a respondent noted that science proceeds *despite* the "subjective" element and that experimental evidence "at the end of the day solves everything." Yet, what happens before this resolution occurs; that is, in the approach one takes to gathering evidence? Similarly, what happens when the evidence gathered is insufficient or contrary to expectations, such that one's efforts meet with as much failure as success? Then the subjective element might appear in a number of ways. The following quotation suggests that the subjective element is in a way contained and translated into the meaning and significance of the experimental result[18]:

> In science, the researcher...questions his object while the object demands that he accounts for himself. [A]n experimental protocol or an equation, can all shed light on evidence which was up to then unknown and must be named.... As it emerges, this evidence, a product of the [unknown materiality of the world]..., constitutes new edges of the [unknown] while in a twofold [reflexive] moment, the researcher fades and is designated in a presence correlated to the emergence of... [material results] that may be named (Szpirko, 2000, p. 16).

The following discussion examines this difficult cusp that marks the interaction of person as scientist with the person as gendered in relation to the issue of experimental failure. As noted, the gender–science relation is not a simple and direct one and requires interpretation that looks beyond

---

[18] Also of interest, she saw more gender discrimination than did a comparable White sample from the same institution; see Malone et al. (2005).

what is said clearly and obviously about gender in the laboratory to what is said figuratively or to what is not said.

"Perfectionists" and "Floozies"

The (light-skinned) biracial participant previously mentioned seemed on the whole to view gender as more salient than race in her lab.[19] In this passage, she remarks on her and other women's "self-critical" stance in doing experiments; her observations provide a springboard for thinking about ways women interact with each other and with their experimental objects:

SP11: *Those are the thought processes. I think it's, I think it's any first-year grad student coming in there, they just don't know what to make of research really.*

I: Uh huh. So, you don't think it matters, you could –

SP11: *Yeah I think it, I think it's amplified for women . . . um, I think women in general are perfectionists and we really like to do stuff well, whereas men are, I don't know what the word is, they'll try whatever and – yeah, I mean, that's my perception, um, I think.*

I: Do you think they feel, sometimes they call it confidence.

SP11: *I'm not sure it's confidence* [Interviewer suggests authorization – participant does not take up prompt but continues]: *Yeah. I think women are raised um, without that motivation, whereas men, you know, they're always being given – you can do this, you can try this, you know, go ahead and do it, whereas women are like, they're raised, Oh don't do that, you'll look bad or um, you know, don't wear that to school, you'll look like a floozy or whatever, you know?*

I: So, you feel like maybe women are raised in a way that they're a little more self-conscious?

SP11: *Yes. Yeah, I think women are raised to be more reserved and self-conscious.* [picks up prompt] *and um, more self-critical.*

I: Uh huh. And so that plays out?

SP11: *Yeah. Absolutely, absolutely. Yeah. I mean, I sit with mostly women in our office, we all have the same perspective, you know, we're all afraid of an, an experiment failing, whereas the guy that sits in the row next to me, he'll try anything* [laughs] *and he doesn't care – he just moves on, you know and because he tries so much stuff, so much stuff works. You know? You have to try a hundred things to get anything to work, you know and if I only try two things, I don't have as much of a chance of getting something to work.*

---

[19] This characterization is also found in the literature. See Conefrey (1997); Etzkowitz et al. (2000); and Sonnert and Holton (1995).

SP11 characterizes herself and other women in the laboratory as perfectionists,[20] whereas a male participant, A10, describes his own working style in a way that sounds much more cavalier:

> *Uh, you know, I know myself, I pay attention to my experiments and I have very organized experiments, but as far as a detail kinda person, I'm not, you know, um, I mean, I guess I, I don't know how to explain it as well, but uh, if, if there's a protocol that works – I won't start again from scratch, to end up with the same conclusion, just so that I know, right –. If it works, um, I'll use it, and go with it – and it's okay, you know, when it doesn't work, then I do stuff. . . . [I] [d]on't care if it's optimal* [2003–10–24-A-i-A10].

A typical way of accounting for different styles of working evidenced by men and women is to suggest that women lack confidence, that they fear failure because of such a lack. Yet SP11 says it is not confidence, but a matter of being self-critical and a perfectionist. Although being a "perfectionist" is generally regarded as an individual attribute, it is more adequately understood as an interactional resource that she draws on as she positions herself within science: "*Am I a good enough experimenter?*" (see McIlwee & Robinson, 1992). Later in the interview she explicitly links women being "*more self-critical*" with being "*all afraid of an experiment failing.*"

It is interesting that SP11 makes an association between women's *perfectionist* tendencies (which implicate a fear of failure) with concern about looking like a *floozy* and how one looks to others at school ("*don't wear that to school, you'll look like a floozy or whatever, you know?*"). The potential for failed experiments overlaps with the potential for being perceived as a floozy, and dressing in an unproblematic manner is associated with looking bad in one's research. A26, a postdoc, also refers to clothes and women precisely in terms of how women handle failure. To contextualize her remarks about clothes, we must note other remarks that A26 makes about women in science. She makes an observation about how women do science and also

---

[20] Although it did not come up explicitly in our study, it is interesting to note that in Sonnert and Holton's (1995) *Who Succeeds in Science*, female scientists spoke of the tendency of male colleagues to publish very similar work over and over or to submit work that is not really complete. By contrast, they viewed themselves as preferring to submit more carefully crafted and comprehensive research publications. We are making no attributions about the inner states of men whose gender enactments are inflected in ways just as particular as those of women who practice science (see footnote 15). Some refer to a masculine style of domination in science, drawing on research in primatology, biology, and from historical records (Haraway, 1991; Keller, 1996), but that is not all that is at stake. It might simply be that, from the masculine position, materiality is presumed to be malleable to one's interests (eventually), such that one needs to just try and try again. One can trust the system or the process, which implies that one fits in better within it.

notes that her own approach is different from that of most women. In lieu of knowing their "*absolute value,*" women are competitive:

> *I've seen with women that sometimes it's personal, right, so and one women gets it and other doesn't, they ask, why did she get it* [e.g., a grant]*? When really that's not the question, you control only what you do and did to get the job, what the other person did was really irrelevant – ah, and maybe that's the difference, women are always comparing themselves among each other and men are always looking at the absolute values of what they are* [2003–10–20-A-i-A26].

Elsewhere in the interview A26 conveys excitement over objective results: the data and the "*thrill of the graph.*" She expresses a love of its timelessness – just as she admires men's sense of their own absolute value. Indeed, she prefers male labs. The point of contrast between women and men in this account, as with that of SP11, is women's concern with appearance: being seen as attractive or being an (attractive) object for another. A26 notes,

> *And you know they say that women dress up for other women, maybe that's a part of it; it's all a matter of competition. I personally don't like that – not that I'm not a woman or have the same faults as women. It's, I just see it that way – what I accomplish is based on what I can do* [2003–10–20-A-i-A26].

Both narratives associate the behavior of the object of the experiment (*might not work/might work out*) with the emergence of the subject as a being of a certain sort: A failed experiment equals being a floozy, or women's position in science is stabilized through personal competition, which is then cast in terms of appearance. A8, another postdoc, also generalizes about a special competition between women, although, as we note later, she also feels that women have a particular way of dealing with science that aligns them as well:

> *I mean, when I was in the all-man lab, I was the only woman and I didn't have to deal with the competition women seem to have that's exclusive between other women. There's like another level of competition that you have only with other women. I think, I think some women, especially in an environment where they're around men all the time?* [I: Uh huh.] *They come up with their own self-identity, and part of their self-identity, part of what's special about them is the fact that they're a woman, right?* [2004–1–21-A-i-A8].

Anxiety stemming from the possibility of failure in relation to experimental outcomes might function differently in the self-positioning of men and women, especially those appropriating some traditional notions of femininity, as suggested by the two accounts provided earlier. One particularly striking component of SP11's narrative is her linking of failing at experimentation to being a floozy, a clearly gendered descriptor. Her view of women as perfectionistic – or what A10 calls "*succumbing to the detail urge*" in describing his lab mates (predominantly women) – contrasts with A10's self-described behavior as having two or more projects going at any one time, so that when something fails at least there is a back-up. We also see that he is not so preoccupied with being optimal. He does what he does to get it done.[21]

PERSONALIZING RESEARCH

The style in which one deals with failure is only one aspect of gender enactment in science practice. Gender enactments manifest themselves in a variety of ways, not all of which are negative. For example, D7, who identifies as lesbian, describes how science offers a respite from family demands to be a traditional woman (she is from India):

> I never told them, because, see they are elderly and they are like really innocent people and a little bit of disturbance would cause a, cause a big um, big tension to them. Already I was going out of the way, like, it is not allowed, it is not expected for girl to take science. And to study and my dad really didn't want me to study, he wanted me to be homely [sic] girl and find a man and get married and all that stuff [2004–11–17–D-i-D7].

D7 goes on later in the interview to say that science allows her to be a lesbian and dress in any way she wants:

> And if, suppose I don't want to get married or, it's like really weird all the way – if I'm not a bisexual person, I'm homo. . . . This is too big thing [for my parents] – like studying science was too big a shock for them. So, I don't think they will be, like wearing this kind of clothes, and thinking very [tape becomes inaudible].

So here science is again juxtaposed to traditional femininity, but from a much different angle. In science, D7 is free not to be a traditional woman.

For A26, women do not know their absolute value and thus are both more personal and more competitive; yet the competition is couched in terms of what one *wears*. Science offers A26 a place to escape from competition with women, and experimentation provides an encounter that she reports as thrilling, eternal, and timeless. Yet there is something intriguing about how

---

[21] This conforms with existing research (e.g., Etzkowitz et al., 2000; Rosser, 1997).

A26 organizes the ideals of science and her own femininity. Her description of women in science ambiguously locates her own positioning as a woman: *"not that I'm not a woman or have the same faults as women."* This is an interesting "unprovoked denial" (Fink, 2007).

In our view, this is a subjective moment in which A26 is attempting to integrate aspects of her identity and her personal trajectory within science practices. Despite the positive framing of science as offering a respite from competition with women, this competition with women has a negative impact on her experience, as exemplified in the following interview excerpt:

> A26: *I actually thrive in a predominantly male environment just because the communication is slightly different. Just empirically . . .*
> I: I'm interviewing so you gotta tell me what you mean . . .
> A26: *It's different because the example that I always take is you know two men, if one guy punches the other, the guy who got punched may ultimately forgive that guy and they become friends or they may punch each other and they are still friends. It was an incident, they addressed it, and they resolved it, and it was gone. Women, and I put myself in that category as well, if I got punched I am not likely to forgive that person ever. Clearly I am never going to be their friend.*

[Both interviewer and participant laugh.]

> A26: *Communication is a little more upfront and you know in industry when men vie for a position and they don't get it they're not mad at that person, they may be bummed about not getting it – but they tried and they didn't get it and they move on.*

Similarly, A22, another woman who had worked in a predominantly male engineering lab in industry, speaks of her perception of how another woman acted just *"like one of the boys."* Although she could not quite put her finger on what it meant to be just like the boys, A22 talks about a style of interaction and competitiveness (staying late, etc.) and that whatever this *just like the boys* was, she could not do it. This becomes a place of negotiation for her:

> *No, but honestly, I mean, I can't tell you why, I think, uh, anymore, why I thought she was acting like one of the guys – trying to be like one of the boys, but it had to do with, like, some of the hours, she kept some of the things she tried to do. I think it may have even been like going to lunch at somewhere I wouldn't go to lu – I don't remember, but I just remember thinking that and – and looking at how her interactions were with everyone else, and how that worked. And consciously saying to myself that's not going to work for me I can't do that* [2003–12–11-i-A-A22].

One can see in this passage that A22 is negotiating her relationship with her lab mates and how to fit in. Here the issue seems to be social (i.e., going to lunch), but when she speaks of the other woman acting in a competitive manner with the men, this issue enters into not just how one interacts with others but how one deals with the project of engineering itself.

The issue of competitiveness with women becomes more important when juxtaposed to the perceived relational inattentiveness of male colleagues. According to four female participants, men go about their business oblivious to the ways in which their activities affect other researchers. Here is one example:

> A8: *Yeah, when I worked in that lab, and I loved the guys in that lab, I mean I really did like them, even though, like after a while I was a little, I needed some female companionship, but um, but I did, I would get frustrated, frustrated is the right word.... And so for another example for you, there was a piece of equipment, right, that I was starting to work on, and there was, there was one piece of equipment that we were having to negotiate time schedules to share. So, what would happen was, we build a whole new apparatus, right? New everything. All of the pieces that go together. So, we built this whole new one. So, now we have two copies of the same thing. So, one was for me to work on. So, I started working on it. So, I started experiments. I go in one day, it's all taken apart. [laughs] I'm like, "What happened?! Why is it all apart?!" And I asked and they go, "Oh well, we needed this little thing here and so we like, took it." And I'm like, "OH! But that's mine to work on!" And they didn't even talk to me, I didn't even get email, "Sorry, we had to take this." I didn't get like – no communication! I didn't get...*
>
> I: And it wasn't... You didn't feel like they took it apart because, "I was a woman, and they don't care about my work"?
>
> A8: *No, no, no. I don't think it was about me being a woman. I think it was about... this is how they are. Like, I think if they had done that to another man, like for example, if I did that to A10's* [a man in her current lab] *stuff, A10 would go, "Who's the male in our girl lab?" Like this, because he does it to us all the time. And, and he does it because he doesn't even think. And he goes – he would just not care. He would go, "Oh you did? Oh, okay, huh." Like he would just say that and if I did that to another female in our lab? I mean, I never would. It's just overstepping boundaries, right? I would... I would email them and I would say, "I'm so sorry there was nothing else I could have done," or I would've, of course, at first, try to go to them and say like, "I really need to do this, you know can we work this out?" Because it's a boundary thing, right? If someone is working on that piece of equipment, even though it belongs to the whole lab, right? Clearly there's one person that's like working on that right now, and so, we would respect that, and*

then communicate about it. . . . It might be that women . . . it might be that women, maybe . . . Is it possible that we're territorial? I mean, I can speak for myself, and I can be territorial. And I definitely personalize research. I mean, definitely personalize research and I'm territorial. I can't speak for all women, but I am. [laughs] *Right. Which is why when someone takes the piece of equipment I'm working on, I get upset, because, I mean, that's my territory temporarily, right? And I personalized it, I said, "Look what you did! That affects ME," because it affects my research, so it affects me. And from his point of view, it's the whole lab and you know, the whole lab works together, and the lab needed this, and so, he did it. Right?* [2004–1–12-i-A-A8].

Despite her discomfort with the word "*territorial*," A8 insists that she took her work more personally, here meaning territorially, than her male lab mates. In trying to unpack what "personal" meant, A8 circles around the issue; it is difficult for her to articulate. As the interview continued, she strays from speaking about being territorial to speaking of a different way of talking about scientific ideas to different ways of relating to research. She associates whatever was particular about her style of doing science with being with a female colleague. Note that A8 does not feel any hostility from the men in her lab; it is not gender discrimination. She likes the men, but she sees a difference.

Throughout the interview, A8 reports a certain frustration with men's indifference to the personal aspects of research at any number of levels, from how they talk about science to how they order supplies to their relationship to instrumentation. In sum, A8 contrasts women to men but says that a man who would seem not to care about another's experiment would not be motivated by selfishness or hostility. Rather, it would be because the men relate to the system differently. The lab, the bigger system of which the man is a part, is what matters to the men in A8's account.

A26 also suggests that being impersonal is more typical of men in labs. As we noted, she expresses a preference for male labs. Her description of a male lab conforms to rather classic descriptions of what women say about what they generally do not like about male research labs (Sonnert & Holton, 1995):

I was the only female grad student and what I really liked about that environment was that . . . we actually had a very small space, we all actually sat next to each other and literally this close and that fosters a lot of communication when we were talking about ideas. . . . [The men] *would not necessarily help you but they would tell you that you are doing it wrong* [laughter] *and then you know at least that gives you the knowledge to work on and they were not mean people, I mean you could ask them for help. They kind of gave you the one-word answers, is this right, no – you're not going to expand on that? No actually you work it out – that was an environment where it was mostly men and, although it was not personally that friendly a lab, it was very productive.*

As A8 has noted, a woman may negotiate her relationship to her work differently than men would. For A8 and other women, their work is more personal: It is "mine." It belongs to the experimenter in a different way. It is more grounded in a personal system of relationships, which for A26 leads to competition between women, for SP11 more anxiety, and for A8 thoughtless and pilfering male colleagues.

In A8's view, women talk about science differently in comparison to the narrower way in which men communicate about it. Yet this different way of speaking may be gendered with positive or negative connotations. In this vignette, taped as part of field observations of a lab meeting, A8 is rapidly presenting and answering an increasing number of questions from an audience made up of lab mates and her principal investigator (PI):

A8: *So what I am trying to do is that I am trying to be more quantitative, I've looked at these at different stages. So what I am doing now is that I am just looking at these first four days of differentiation. I am looking at what is happening with these markers.*

A8: *Yeah and you're going to see this, I'm going to show you what happens but basically what's looks like is happening is that I have two populations of cells at this point and I can see that by seeing two distinct peaks on FACS gram and I'll show that to you. Are you guys ready?*

A13 (PI): *No.*

A8: *And so it's the first slide.*

A13 (PI): *I'm not ready.*

A8: *Oh, I'm sorry –. What else, what else can I help you with?*

A5: *She sounds like a waitress.*

[GENERAL LAUGHTER; everyone is laughing loudly.]

PI then asks a question that is research oriented [2003–11–07–m–A–i–A8].

Interestingly, several women from Lab A attended a local presentation of our research at a women's studies lecture in which we described this exchange as implicating gender in the lab and as showing a counter-position to A8's identity as a scientist in this presentation (i.e., a waitress). When we interviewed one of these women, A5, shortly after our talk, she vigorously disagreed with our interpretation, saying that she would have called a man a waiter had he acted the same way. Nonetheless, A5 did agree with part of what we were saying. She noted that her remark to A8 was meant to chastise A8. She was trying to *"police"* A8 because A8 was wandering off the scientific trail: A8 had become too personal. She had lost the boundary between science and the personal in that she was taking criticisms too personally. A5 was trying to deal with A8's taking things too personally by making a joke about her sounding like a waitress.

A5's viewpoint about the lab meeting was quite different from that of A8. A5 described a lab meeting where there was increasing tension over A8 being so personal and A5's remark blew off steam and got A8 back on track as a scientist. After much additional conversation with the interviewer, A5 gave the following broader explanation of her actions:

> *Yeah that is, that is what I was doing. Because, but the only way to say can you switch back to scientific mode now was to completely break the track that we were on, like it was drifting further and further away from scientific mode, you know. There is actually something interesting in when it comes to science. It's a skill that scientists have to learn and, ah, some people do better with it than others and that's talking about the science, talking about the experiments, but it's not talking about the person – does that make sense??... So if I am talking about and it can drift into a more personal criticism but usually it's not, especially in a lab meeting context – only in a superheated conference discussion you, with foes of opposing, you know, whatever – it can drift into personal attacks. So if I am talking about the experiments that I did and blah blah blah, yes I did them and it was my ideas but what we are talking about is the science and the experiments. We're not talking about you know, do you think the experiments are designed well, not do you think I am good at designing experiments. And these are two different things and so people tend to drift, some people are better than others. And there is a wide range within our lab of people who are comfortable with that skill. So that comes up A LOT* [2003–12–13-i-A-A5].

Here we see that, although A5 denied that the waitress joke had any gendered connotations, she aligned it with taking research too personally. A8 characterized this personal touch as feminine, even if it causes some problems with men. In contrast, for A5 (and A26), it is simply about science – science lets one be less personal. For A26, men are better at this sort of scientific stance. For A5, there are other ways to bring in gendered experience, as we see later.

By contrast, A8 barely remembered the incident, but did say that there was no tension and that she was easily taking in all the advice and criticisms. She remembered very clearly saying, *"Can I help you with anything else?"* Yet she attributed the remark about being a waitress to the principal investigator, who joked about sex and gender quite affably during lab meetings. She has no recollection that it was made by A5 or a woman. Overall, then, this issue of being *personal qua gendered feminine* lurks within science, from the style of presenting data to one's attachment to experimental results, to how one experiences the instruments one builds, to how one imagines the costs of experimental failure. As we said at the beginning, one can see it emergent even in descriptions of scientific work, but our sample is not sufficient to fully characterize gender effects within labs for the sample that we have researched.

CONCLUDING REMARKS

In the face of both the rigors of science practice and the failures that attend any research efforts, each scientist develops a style of working and communicating that sustains personal motivation. This style or "personal stamp" in laboratory work is gendered to some degree. If gender is a "master" identification, meaning that it is inescapable, each scientist's personal history is in part a gendered history. One might say that subjective formations (the stamp of particularity) are crafted through gendered histories in each scientist's effort to render science more bearable, more exciting, and more in sync with other identifications, so that there is compensation for the sacrifices required by science. Importantly, however, gender is enacted heterogeneously. There are many women who act like men or who are conflicted about their position as women within science. There are some who have not successfully mastered this position, others who progress successfully but only at some emotional cost, and still others who have found a way to enact it advantageously.

Given the heterogeneous ways in which gender is enacted, analysis of particularity – of the person – is a necessary first step in understanding the role of gender in science. In the particular activity of the scientist, the effects of gender and the demands of science make their compromises and indicate future points of negotiation. Indeed, it is only such analysis that would enable either the claim that science is gender blind or that some aspects of science are irreconcilable with gender, whether science offers an inhospitable atmosphere or an inherently biased structure. Yet this is a level of inquiry at which much data remain to be gathered, both in science studies and in feminist research.

7

The Learning Person

Learning is among the first topics covered in introductory psychology text-books, typically after chapters on sensation, perception, and neuroscience. This prominent positioning reflects in part its importance in the greater history of psychology as a laboratory discipline. Although most robustly embraced as essential to psychology's concerns by behaviorists (Skinner, 1938, 1953; Watson, 1913), concern with learning overlaps with the shift to analysis of representations and memory mechanisms central to the cognitive revolution (Baars, 1986). In application, learning is the obvious core of educational psychology, but implicitly it is also the very machinery of clinical intervention, being a matter of "relatively permanent change in behavior or mental processes due to experience" (Huffman, 2010, p. 203). This definition is taken from an introductory psychology textbook and is typical of those given for learning in general psychology textbooks.

The breadth invited by the definition of learning as "relatively perma-nent change due to experience" ensures its analysis at multiple levels of organization, from synapse to culture. Learning is thus central to our focus on the acting person, a unit of analysis we consider to reflect multiple levels of process and engagement in various interactive systems – despite the fact that we must by necessity narrow the focus to different levels at different times for the sake of imposing order and seeking clarity.

Thus in contrast to the convention honored by most general psychology texts of explicitly focusing on learning early in the sequence of chapters, learning is here the last topic prior to our overall conclusions. It is fitting that the learning person is the culmination of our examination of acting persons. Frontier laboratory work inextricably implicates learning, working, and innovating. One learns through work and one innovates through working and learning. This is not a cycle but more a cross-hatch of co-implicated activities. Failure is an ever-present catalyst to this process of working, learning, and innovating. Although much is repetitive in lab work, a greater portion involves creating new devices, assays, or detection strategies that often fail and then responding to such an impasse. The learning person solves problems, feels emotion, and undergoes identity transformations. Thus it is fitting to focus on the learner and his or her learning journey in the research lab as an integration of previously explored themes.

Our approach to the topic of learning in science practice is to present a case study of a single researcher, through successive interviews over the course of a year that began shortly after her entry into Lab A as a new master's student with a mechanical engineering background. The case study analysis enables us to further address one of our central objectives: to consider how a focus on the acting person in the normative contexts of science practice informs both science and psychology. Because the whole sequence would be too lengthy, we focus on the first four interviews that track her early learning and on an interview conducted a few months later, shortly before she decided to enter the PhD program.

THE INTERDISCIPLINARY RESEARCH LABORATORY AS DISCURSIVE PROBLEM-SPACE

To understand the interconnections between learning and other dimensions of human activity, we conceptualize the laboratory as a *discursive problem-space* that *positions* researchers to *act* in ways that promote *learning* through making *strategic contributions* to the work of the lab. By using the term *discursive* we mean to emphasize both that communicative interactions and community are central to laboratory processes and that these processes are of extraordinary complexity. As we emphasized in the chapters on the problem-solving person, the feeling person, and the positioning person, the laboratory community includes not only other persons but also the entities and technological artifacts of laboratory practice, by virtue of the relational strategies researchers engage in with respect to these entities, particularly cells. Moreover, the cells, as living entities, are equipped with a repertoire of action possibilities of their own, including dying. The actions of cells are cooperatively mobilized with researcher actions toward various problem-solving goals, as we suggest with our notion of cognitive partnering with cells (see Chapter 3). Therefore, cells and the simulation devices designed to interface with them (model-systems) are part of the discursive space of the researcher in a biomedical engineering laboratory, as are all of the other researchers and the normative traditions and conventions of the field and of science. Moreover, by calling it a discursive problem-*space*, we emphasize that the physical arrangement of the laboratory is vitally important to the forms of communication enabled and supported within it.

The discursive problem-space of the laboratory offers what we have termed an *agentive* learning environment. *Agentive* as an adjective implies two concepts: *agent* and *agency*. The notion of *agent* emphasizes the learner as empowered to form relationships with humans and artifacts, achieved through changing types of participation in the work of the lab. As an agent, one *acts in* the discursive space and *acts with* other researchers, devices, equipment, and instruments. Agentive in this sense implies the person/ learner who is characterized by relationships to the discursive problem-space (in contrast to the individual, who would be characterized essentially by separation from other participants). It is these growing and changing relationships that the *agent* dimension of agentive is meant to capture.

Agentive as related to *agency* is a bit more complicated. Over time, learners achieve a level in which they *act through* cells or a flow loop or a vascular construct model-system as cognitive partners, as we saw in Chapter 3 on problem solving. At this more advanced stage, the researchers cultivate and evoke certain artifacts as mediators and cognitive partners in their work, thus assigning agency to that partner. The *agency* aspect of the agentive, then, emphasizes *human* agents who are "authorized" to privilege other entities, human and nonhuman, as agents in their research (Newstetter, Kurz-Milcke, & Nersessian, 2004). To achieve this level of understanding, learners work through phases in their engagement with lab members and artifacts. The lab as a learning ecology supports and encourages this growing understanding of the partnerships that are necessary to accomplish the immediate work of the lab and to reach long-term science goals. The learners, on their journey in an agentive learning environment, have the opportunity to *act as to try on* various positions/identities as we see in the following case study and, in doing so, to identify with and become a member of that science community.

The discursive problem-space positions each researcher to make strategic contributions in two broad categories of action: (1) sense-making and (2) identity formation. These may also be considered two facets of learning taking place in the laboratory. Importantly, the discursive problem-space is produced by the agentive contributions of persons acting within it, but it is also *productive of* agency, in the sense that new possibilities for action (reasoning, problem solving, feeling, and forming identities) are established. Learning can be understood as the enactment of these possibilities, as we attempt to demonstrate in this chapter.

The interrelations among all of the categories examined in this text can be illustrated in any given interview. Yet the subtle, delicate ways in which the connections among categories show themselves in speech become even more compelling when we consider a series of interviews conducted with a single researcher over time. We have conducted detailed "case studies" for a small number of researchers, with whom we carried out a series of interviews at quite regular intervals to follow their sense of their learning path. In addition to helping us refine our codes and conceptual categories, the case studies enable us to track developments in learning, modeling practices, and identity transformations as we explore the forms and conditions of the processes of one person's learning. Transformations in identity in relation to other researchers and laboratory artifacts and the interfacing of affective and motivational factors with cognitive practices are important components of these processes.

CASE STUDY – A22 IN LAB A

Progress interviews with A22 began several weeks after she joined Lab A and continued at approximately 2-week intervals for a year. At the time of the first interview, she was a master's student in mechanical engineering (and uncertain whether she would transfer to the PhD program) and in her late twenties. She had earned an undergraduate degree in mechanical engineering from the same university. Before returning to school for an advanced degree, she had worked in industry on fuel injection systems for diesel engines, but then decided she "*didn't care about engines.*" She said that her motivation for returning to school was that "*I wanted to feel I was making a difference.*" This was her first research laboratory experience and her first foray into biology. A22 described an early fondness for mechanics, cars, building things, and "*doing hands-on kinds of work.*" This led to her pursuit of engineering, but she was disappointed to learn that most engineering jobs involve sitting behind desks rather than working in the hands-on way she had anticipated. In a later interview [2003–12–11-A-i-A22] she affirmed that most of her colleagues in school and in her first working environments had been men. In contrast, Lab A was staffed mostly by female researchers at the time of this sequence of interviews. She noted, "*I'm not used to working with this many women. And when I've worked with women before, they've all worked in very different fields. This is the first time where I'm working with women who are my co-workers, you know, like in my same lab.*" Also new to A22 was that she was working closely with an African American colleague for what appeared to be the first time, though she noted this fact without any discussion.

   The section headings that follow are direct quotes from our learner, A22, in each interview under review. Each captures her changing forms of participation in the life of the lab at the time of the interview and thus serves to chronicle her learning trajectory.

"I've Got My Own Cells" (10–10–2002)

In Lab A, cell culturing and maintenance serve as the starting points for lab learning. As one senior graduate student commented, learning to culture cells is "*baseline to everything.*" For A22, a mechanical engineer by training, understanding cells' needs, processes, reactions, and possibilities is something that will occur over time. At the beginning, she experiences the cells in culture as objects that can be manipulated in various ways. In our first interview, in response to the question, "So what are you doing?" she states that she is "*learning cell culture.*" She follows this reference to *learning* with a description of the work she is *doing*: "*So – I've got my own cells that I'm feeding and I'm splitting, and I probably will freeze.*" She affirms the close association between learning and *working* in a series of responses to the interviewer's query about from whom she has been learning these tasks:

> Well, A5 [a senior graduate student] – we sat down and she worked with me. She showed me – we got cells – we had six different groups. She showed me like three of the – with three of the cell cultures and then she watched me and corrected me as I did the other three. And then I just came back and kept doing it and if I had any questions, I could ask.... And with A7 [another graduate student], mostly I've just been watching her work, so far. And then today, this morning I'm going to be doing trypsinizing where basically A10's [another graduate student] going to tell me what I need to do and I'm just going to watch. I mean he's going to watch me and make sure that I do it correctly.

This interview excerpt describes the apprenticeship relationship as enacted in both labs we investigated. First, the learner watches as a complex task (work) is demonstrated several times. When she feels ready, she does the task herself with coaching from the observing mentor. Finally she undertakes the task solo and asks for assistance only when needed. This changing dyadic interaction not only constitutes the learning configuration needed to master complex techniques, but it also serves as the foundation for forming relationships with other people in the lab. As the researchers learn to work with cells, they learn the value of working/learning from lab members. Apprenticeship enactments entail identifying next learning opportunities – strategic learning through watching, doing with, and then doing alone – and elucidate how Lave's notion of legitimate peripheral participation (Lave & Wenger, 1991) unfolds in these research communities: "A person's intentions to learn are engaged and the meaning of learning is configured through the process of becoming a full participant in the sociocultural practice" (p. 29). Through the initial apprenticeships, the learner becomes cognizant of and a participant in the cognitive, investigative, and interactional practices of the lab (see Chapter 3). The first steps toward full lab participation are supported through the need to culture cells and build constructs with them. These activities, which will soon become routine and unremarkable to the learner, not only afford first forms of membership but also start the process of acting in the lab with artifacts that are central to lab life.

Noteworthy in this excerpt is A22's excitement that the cells she has been working with are her *own*. She is developing an identity as a caretaker/owner of the cells and as a lab member (see Chapter 5). Having one's "*own cells*" is a significant marker of a changing relationship to the lab, the work in the lab, and lab members. Before they have ownership, lab newcomers tend other people's cells; they are positioned as "custodial workers" whose rights and duties are prescribed by the cell owners, akin to a lab technician. With their own cells, the new lab members are positioned with a different set of available actions, including developing an affective relationship with them (see Chapter 4).

At this stage for the initiate, things seem relatively simple and straightforward because the cells seem to be malleable and compliant; the cells are something you can get and own and do things to. As we noted before, a graduate student in this lab was overheard telling a newcomer: "*Think of them as children or pets.*" However, such metaphors fail to capture the full challenges of working with cells. In a later interview that is not part of this series, A22 characterizes her first cell encounters as "*complicated*" because "*you have to focus on the moment*" and you have to "*think about things you can't see.*" Cell invisibility and the recalcitrance of the biological side of things offer challenges that differ from those faced by engineers. A minute error in hand positioning under the hood can ruin an experiment and cost several weeks of work. Although not "seeing" in engineering work is often costly over time, not "seeing" in the lab environment is immediately damaging – a first lesson with the cells as potential partners.

In this first interview, when asked where she thinks her research may be headed, A22 is uncertain but clearly co-implicates learning, working, and doing or acting in the space of the lab with lab artifacts:

> I still need to learn more about it, but I'll be working with the fabrication
> of constructs with the possibility of either looking at different methods of
> fabricating constructs or – uh–making – basically a construct within a

construct. But before I can get there and really understand it I have to really learn cell culture and how to do a construct and I have to learn cell culture before I get there.

When asked by the interviewer to further explain the phrase *"construct within a construct,"* she describes it as

a construct with an inner layer, that is, the collagen with smooth muscle cell and then an external layer which is a collagen, which is simply collagen, without muscle – without cells. I guess that A11 actually did something similar but in reverse order, with the smooth muscle cells on the outside.

In this explanation she exhibits a model-like understanding of the construct that is both engineering and biological. The reference to reverse order suggests not only structural modeling of the construct within a construct but also a temporal or sequential order of the process and a manipulation of the representation by imagining the order reversing.

Summary: In this first interview, A22 is finding her way – sense-making – in the world of biology through owning and culturing her own cells. Although tentative about her future project, she is clearly enjoying her first lab experience because of the fluid mentoring practices of lab members whom she characterizes as *"great"* and *"helpful."* In particular, A5 has *"gone out of her way,"* offering to help *"without having to ask her."* Indeed, she notes that she has found everyone in the lab willing (*"incredibly willing"*) to help and to meet with her as needed. She feels well supported in her learning and fully engaged.

"I Finally Found a Use for All Those Cells" (10–23–2002)

As alluded to in the previous interview, once newcomers have mastered cell culturing, learning to build "constructs" or the bio-substitutes foundational to lab research comes next. Two weeks after her first interview, the second interview begins with this exchange:

I: What's it been like in here lately?

A22: *Fine. I learned how to make my – make blood vessel constructs. I did that this weekend. I finally found a use for all those cells I was culturing. That was interesting.*

Here we see the sense-making that has occurred for A22 as the lab technique of cell culturing has taken on new meaning: Cells are not just objects to feed and sort but are also the building blocks for constructs, objects that she not only acts on but acts with, in concert toward a greater goal. A22 goes on to describe the actual process of construction:

> *Well first you have to – figure out what cells you want to use, you use trypsin and you take them off the plate and then suspend them and you put them in media – and count them – how many cells you have and based on that – how many constructs you have. There's spreadsheet that someone came up with that you determine how much solution you need – and then I have a nice little cheat sheet to go by which is much more helpful. You use these tubes, and you put in, in the cells suspended in the media, and oh yeah, you have to add – like – some collagen and some other factors to get it to gel, and they have the little 3-mm glass rods that they've made that I will, I guess learn to make at some point, with little stoppers at each end that they use to put in there and the cells bind that and kind of gel to a nice little construct form. So when you put everything you need in there along with your little glass rod you put them in the incubator. You let them sit there for an hour until they gel and then after that you get a dish reader and put some media in them and transfer them into that and then they need to compact so and the – like the next day you come back and you make sure you cut the construct from the ends of the little stoppers so that – because you want them to bind in the middle so that they form so they line up and they form fibrin that are run radially, which is the way they do in your blood vessels rather than from end to end which is what they will do if you allow them to be constrained at either end.*

Here our learner interacts with the construct and its wetness as the "latex-gloved manipulator." She has learned how to culture the cells and how to build constructs with a spreadsheet that "*someone came up with*" and "*a little cheat sheet*" of her own. We see her here as an agent and a problem solver in both using what others in the community have developed and in crafting her own form of assistance in mastering this many-stepped task. In this learning phase, the construct is something she creates from components that have certain dimensions using highly specified steps. It is the product of her manual work under the hood where she does things to and with the cells, tubes, media, collagen, rods, and stoppers. As with the cells, her preliminary understandings are mediated through manipulating objects or acting with them. She will build these constructs repeatedly for practice and later for the experiments she will perform. These interactions and activities characterize her first relationship with the constructs, a relationship enacted through physically manipulating the lab materials, through working. As an engineer and designer (her identity at this point), these manual actions align with her stated desire for hands-on work. When asked why the construct gels, however, she responds, *"It's due to the additive in there. It causes them to kind of come together. Yeah, I can't give you a really good explanation yet."* A22 falters when asked to explain chemical and biological processes undergirding the device creation. The development of a conceptual model of the deeper biological processes or an identity that incorporates a biological dimension is not yet evident.

As a result of making her own constructs, she has enhanced her ability to think more cogently about a future project – about formulating her own research problem. A month or so prior to this interview, A22 had started discussing a project with the lab PI but she reflects, *"At the time I didn't know anything, so I didn't really know what kind of questions to ask."* In contrast, anticipating a future meeting with her advisor, she says,

> *Like I know that he discussed fabrication and the possibility of coming up with – a two-layer construct. So those are some things that, um, I'll be discussing with him more now that I've actually made some constructs. Now I have a little bit better idea and I can ask better questions and ask more questions. Although if I have ideas of my own of what I'd like to pursue I can discuss it with him and change the focus.*

Summary: From this interview we see how learning by doing (1) has increased her confidence and engagement (agentive learning), (2) helped her begin to frame a problem, (3) helped her form questions, and (4) facilitated her participation in conversation/problem formulation.

"Keeping My Cells Alive and Happy" (11–06–2002)

This third interview demonstrates increasing understanding of cells and their environmental needs. It begins with a broad question about what A22 has been doing, to which she responds, *"keeping my cells alive* [laughs] *and happy."* This is a first anthropomorphizing of cells as something other than objects to be manipulated, reflecting at the same time agentive participation in the emotion attributed – she is *keeping* them happy. The need to do so is linked to the ever-present potential for failure. She continues to explain that "happiness" derives from *"splitting the cells, passaging them so there are not too many"* – elaborating that if there are too many *"they just grow on top of each other. But you have to feed them more often – and eventually if you have too many, you just can't feed them all."* So we see evidence here of a changing understanding of the cells as mere objects of manipulation to entities that respond to environments in observable – and potentially undesirable – ways, demanding the researcher's attention and response.

At the same time that A22 is developing a deeper understanding of cells and their processes, she holds tight to her identity as an engineer, as evidenced throughout this interview:

> *I am still looking into fabrication of constructs and really it's okay – If I'm going to do this – how am I going to determine what's good? And we were addressing all the different techniques used for mechanically measuring – determining the mechanical properties. One of the things we do is a burst test, which I'd like to learn. Another thing – another option that people do is something similar, but they check the pressure versus the diameter. So that's something I might look into. It should be easy to figure out how to do it, to have everything to do it. Plus there's some publications* [sic!] *out there on how to do it from people who have discussed the possibility of suture strength . . . looking at that. So I don't know.*

We see in this passage a certain level of comfort as A22 moves into engineering accounts of measuring the constructs. She identifies a "burst test" as something "*we use*," signifying her identity alignment with the engineering procedures of the lab and as a test she would like to learn. She moves on to *pressure versus the diameter* and finally *suture strength*, all topics of interest to her. What is most interesting about this passage is the confidence she reveals in figuring out potential determinants of "good" (for evaluating her constructs) and in strategically positioning herself as a researcher who can learn whatever is required for her project. She alludes to strategic learning and constructing a framework or schematic for her research goals in moving forward.

When the interviewer asks for clarification on what the tests would determine, her response, as an engineer, appeals to quantitative representation, but also to norms and standards:

> *Ah, well, if I'm looking at fabrication of constructs I'd want to quantify one method versus another to determine whether or not we, actually one method would be considered better than another.*

The interviewer then inquires whether she has specific methods in mind. She begins by noting that she has "*thoughts*" and that she is "*interested.*" She then acknowledges resources from a working group in Canada, positioning her thoughts on methods as scientifically legitimate by aligning with this group, made up not of engineers but of biologists.

> *There is a group in – a guy up in Canada, I believe, who's really big into biology and they do an interesting method where they create just a layer of cells and they're able to condition the cells so that they're able to line up and*

they line up in a way that would be similar as in the body and they actually roll it into a tube, so that way you've actually got them aligned, not only, not only you have them circumferentially as well.

Also interesting here is that A22 begins to formulate questions and consider the bigger picture (framing) of research conducted outside of Lab A. Her alignment, however, is short-lived. She refines her questions by distancing from the Canadian laboratory's procedure:

So it sounds like an interesting technique. It takes a lot longer than what we do, so I'm interested in looking at it, and apparently they like their method a lot more. But there's some questions [sic] over whether or not the cells actually stay in that nice, aligned pattern. And if so for how long. And so also I'd be interested in seeing if – what the mechanical properties are.

When pressed later to identify what most interests her in terms of potential future work, she responds:

Right now I am interested in learning about the burst pressure and doing that kind of pressure testing. That's something that's really interesting. The suture strength. I know that's been done for other tissue engineered. I don't know that's been done for – it's basically you do a tensile test, you put a suture in, basically and – and see what force it took, and that's something that would be clinically relevant.

In this problem formulation, we see her recasting the creation of a viable blood vessel as an engineering problem comprising forces, burst pressure, and tensile strength (thereby interlocking biology and engineering models) and for the first time mentioning the clinical relevance of the project.

When asked about her interest in burst pressure A22 invokes an engineering approach that allows you to *"get really good data and compare"* as the reason for her interest:

Because to me it makes the most sense for um, comparing to like a – comparing to the human body. You've got blood flowing through the – your veins, and the pressure – it's the same kind of pressure. . . . Um, and, you know, the pulling apart you get really good data and you can definitely compare, . . . The nice thing about the tensile testing is that you don't need nearly as much tissue [laugh], you can make several . . .

She also answers the question about her interest with a statement of professional identity: *"And I like it because – since I'm an ME, technically, I can still feel like I'm kind of, not straying as far – mechanical testing is all ME* [mechanical engineering]." She identifies activities such as flow testing and watching cells and figuring equations for the flow test as things that do not "*stray too far*" from engineering.

Summary: Although we have evidence that A22 is developing a new understanding of the biological aspects of the lab, that cells are not just objects to be manipulated, she is still firmly rooted in her identity as a mechanical engineer. Her search for a future project appears constrained by a desire not to stray too far from her earlier educational roots and training.

"It's Closer to Native Blood Cells" (11/20/2002)

Two weeks later, A22 gives a very different account of the construct and the cells in response to a question regarding her future work with the fabrication of the construct. She had recently met with her advisor about her idea for a project. He is encouraging but has warned her that "*some of the things I was thinking of were good but at the dissertation level.*" Her idea for a master's project is "*to focus on the dual layer construct*" – the "construct within a construct" idea that intrigued her early on.

> I: So are you going to compare – are you going to compare this dual layer to a single layer?
>
> A22: *Our current method – yeah. Because what A11 found is that when he completely dried out the collagen, that even when he put it back into the media and it re-hydrated, that it didn't – as long as he completely dehydrated it – that it didn't lose – that it kept some of its strength. And he thinks that something happens in the dehydration that helps to increase the strength. And so but by having it on the inside, without any food on the inside the – there's no cell growth through that collagen, and you want to have – the smooth muscle cells on the inside so that you can – so that the endothelial cells will be happy. So I'm going to look at – doing– an external collagen, but although probably a more likely scenario is coming up with a method that I could make – that I could dehydrate a tube and then it would be inside the structure rather than sitting on the outside layer. But, I'll try –* [laughs] *I'll try.*

Here we see a very different account of the construct from her previous understanding. First, we now see evidence of a model-based understanding of the construct as a cell environment possessing properties such as strength and as comprising smooth muscle cells, which under appropriate conditions will promote cell growth. The new model fluidly interlocks biology and engineering understandings as needed for conducting research. No longer a compilation of parts (structure), rather, the construct is a mini-system of sorts (behavior and function) that when operating optimally can sustain cell growth and increase strength. Also evident is her intention to create an environment so that the endothelial cells will be happy. Previously she was making the cells happy by feeding and splitting the cultures. Her new understanding is that the environment will have to do the job.

Summary: This new portrayal of the cells as interacting with the environment in a happy way suggests that A22 has moved beyond the simplistic account of the cell as an object to be manipulated and is starting to develop an interactive concept of the cell as reacting positively or negatively to the environment. Additionally, this characterization reveals her understanding of constructs as products of communal activity around a problem – the lack of mechanical strength. A historical trajectory of the lab is constructed around this artifact (device history), giving it meaning through its relationship to the lab members and their different projects. A22 refers to one other lab member, as she alludes to his role in this problem-solving effort in the discussion of dehydration as a possible source of strength. A22 identifies her future work and role using dehydrated tubes on the inside of the construct rather than the outside. The construct here serves as a meaning-creating thread that binds the activities of varied lab participants and etches out a chronological line of research – a hands-on appropriation of the history of the lab's construct. In this account, the construct exists in the plural, as a community resource acting as a "player" in communal problem solving. If pressure is the enemy in vivo as blood pulses through an artery, then constructs are partners in conquering this enemy.

"I Want to Find My Own Niche" (12–13–2002)

Three weeks later, A22 has continued to practice cell and construct techniques, but now distinguishes this activity from "*real work.*" This differentiation suggests that she has reached a stage at which mere hands-on manipulation of cells and constructs under the hood, although necessary, is secondary to another class of work: her own work with the cells and constructs. When asked about her work, her response appeals to emotion:

> One of the things that I'm interested in investigating, that I think would be fun, would be looking at, comparing different types of constructs that I've – you know. I want to stay along the lines of fabrication of constructs, but in order to do that I have find some way to quantify what's better. And uh, we have several different methods, but one thing I think I'm interested maybe in pursuing is – is looking at burst pressure versus diameter changes.

Yet reflecting on this statement about the "fun" of her research, she immediately appeals to community norms and expectations or standards: "*I don't know if it'll be significant enough.*" She then constructs a framework in which identity, positioning, lab history, norms and standards, problem formulation, and affect/motivation are all interwoven in her planning:

> We kind of like this [IDENTITY with lab] *because nobody in the lab is doing it* [POSITIONING within lab], *but, right now, but, we have done the burst pressure before* [HISTORY]. We have all the equipment to do it, so it's not – so it would be something that would be easy to do [PROBLEM FORMULATION/PRAGMATIC]. *I just need to make sure that, I want to* [AFFECT/MOTIVATION] *– there's a lot of literature that from people who've done it for other tissues* [POSITIONING with community beyond lab]. *In fact they may have done similar with blood vessels in another group, so it's something that I'm interested in researching and learning a little more about* [AFFECT/MOTIVATION], *and what to do and deciding whether or not it makes sense* [NORMS/STANDARDS].

This passage demonstrates A22's recently developed ability to frame her own research in relation to the work others are doing in the lab; her growing identity as part of the laboratory community; and her ability to relate to the field in general, the available literature, norms and standards for "significant" work, and her own interests and to integrate these considerations in evaluating what direction of investigation "*makes sense.*" When the interviewer attempts to clarify whether the attraction of this problem comes from the fact that no one else is doing it, A22 affirms this, stating, "*I want to find my own niche.*" This seems to implicate identity but also affect/motivation, a desire to be her own agent in taking control of the direction of her research. She also calls it a "*nice test*" because of its close simulation of bodily conditions; she then elaborates on this characterization by using anthropomorphism while engaging in perspective-taking:

> It really simulates conditions similar to in your body. That an actual blood vessel would see in a body. . . . one of the things that is already known is that when you apply mechanical forces when it's just flowing, across uh, a layer that it helps to increase the mechanical properties as well. . . . And I haven't had a chance to do all that – that's another literature search I want to do. I know they've done it with flow testing, I don't know if they've done it with? kind of pressurizing, maybe I do a little bit, if I do some pressure testing. Can I actually condition <u>my blood vessel</u> [her emphasis] and you know, what's it going to do to my cells, and what's it going to do to the strength? I think that's something that'd be interesting to look at as well.

Here we see evidence that her understanding of the blood vessel is model-based. She identifies gaps in her knowledge and positions her interests in relation to existing research outside the laboratory. In expressing her interests she exhibits developments in her identity within the lab and the larger community, displaying a sense of ownership by taking possession of the blood vessel and the cells. However, she realizes that she will face challenges – strategic learning challenges – in her proposed research because she is an engineer who lacks at this point the biological understanding that will be required for solving her research problem:

> *The biggest thing for me is trying to figure out, it's really trying to figure out how I want to look at, at the fabrication . . . simply because A13 thought it would be great . . . because I'm a mechanical engineer, you know – looking at fabrications seems more mechanically inclined – however, the way this thing works, it is very much biological, so it's not just mechanical. You have to understand the biology and what's going on – because I'm a mechanical engineer. You know, looking at fabrications, seems more mechanically inclined, however, the way this thing works, it is very much biological.*

Here, she recognizes that the actual problem requires more than mechanical engineering; it requires an understanding of biology. It seems that the path that she has chosen is now challenging her to reflect on and even reconsider her identity as just an engineer.

Summary: Overall, we get the impression here of a developing comfort with her work, a sense of the potential for her making a contribution both within the lab and in the wider field. A22 is developing a schematic of various people as resources in the lab, showing greater framing capabilities and displaying a greater sense of strategic and technical knowledge.

"I Wouldn't Say Right Now That I Have a Primary Mentor" (03–02–2003)

To examine how A22's learning is progressing over a longer period of time, we skip ahead now to an interview with A22 conducted approximately 3 months later. At this point she is in her second semester in the laboratory and in the master's program, and she finds herself able to spend additional time in the lab because she is taking only two classes. Yet even with her freer schedule, her projects are taking more time than she had anticipated. She describes herself as feeling "*not really prepared for the amount of time each of the experiments takes,*" but rescinds that complaint with "*but it's good.*" When asked for clarification, A22 frames her agenda as trying to "*dehydrate collagen*" and to use that as a "*support sleeve.*" Her framing of this agenda has a model-like structure:

> *I'm working on getting it to the outside of the, uh, construct so that the smooth muscle cells are on the inside, in the lumen, so that the endothelial cells will, um, attach to them.*

She volunteers that she is pleased that she has been able to accomplish this agenda: "*And I've been able to do that so far. I'm quite pleased.*" A22 then explains that she has been learning how to use the confocal microscope, of which there is only one in the building for the many research projects going on. The complexity of the machine requires a great deal of training so the interviewer wonders why she needs the confocal for her research. This is the first time it has come up.

> *I'm going to be using it for live-dead staining – the reason I'm interested in live-dead is a couple of concerns that I have – it's mainly to make sure my techniques are OK and I don't need to look for other methods um, of either manufacturing, or feeding. One of my concerns with especially adding another layer on the O.D.* [outer dimension] *is "Are the cells on the lumen going to be alive? Are they going to be able to get nutrients?" I'm also concerned because the method I told you about where I'm dehydrating collagen over top of the cells is not a really good environment for the cells.*

Evident in her response is a developing sense of cell ownership and concern for their well-being, which suggest the development of her identity as both as caretaker of the cells and designer of an environment in which they can thrive. This identity development is closely related to an expression of a cognitive partnering (with cells as active participants in her research) and even positioning (cell and researcher).

When asked later in the interview whether A7 is still her primary mentor, A22 replies, "*I wouldn't say right now that I really have a primary mentor.*" She instead identifies A7 and A4 as "*the two people that I really go to for a lot of the work*" because of their experience. She then elaborates on her other working relationships and their specific contributions to her learning:

> *A4 taught me how to use the mechanical testing, A5 has taught me the confocal, I've been talking to A10 about different procedures and how to do things so I've been able to go to pretty much everyone. And uh, A26 has been wonderful. She has a different perspective.*

In her journey, she has relied on these lab partners to teach her techniques and the use of useful equipment; now, however, she has moved beyond a single mentor to being a lab member who uses the human capital of the whole lab – she has moved from an apprentice to an agent in forming her cognitive partnerships. This transformation becomes even more salient when A22 talks through goals for the semester and year ahead. A22 responds that she does not have any concrete goals, but in fact her response suggests considerable strategic thinking. She is clearly formulating and monitoring her own learning goals and desires:

> *No, I don't think I have anything concrete because um, I'm starting to write down on my list of tasks but I don't know that I've really put a timeframe on them. The hardest thing right now is I keep thinking I want to do: I want to learn how to do histology. Because I want to see if my cells are moving into the collagen matrix sleeve, and that's probably the best way to do it. And I know that I want to do burst pressure testing and I know that I want to do mechanical testing and do some live-dead testing. So I've got at least four different things I'm going to look at, and I'm just – I'm realizing that I'm pretty much going to have to dedicate one batch of constructs, you know, for this testing.*

Six months into the lab experience, A22 is imposing order on her problem-solving agenda, breaking it into steps and establishing a timeline:

> *And I'm trying to decide in what order I want to do these things. I think I would like to start doing the histology. I would like to start doing that soon, because what I would like to do is make a batch, test some of them, make a big batch, and test some of them at like seven weeks – not seven weeks, seven days, let them incubate and see what happens. Kind of over about a month time span – just to get an idea of, of how quickly, or if they will proliferate or they will move into the sleeve. Uh, so, so I'd like to get that started because I know I'm going to have to have a lot of incubating time.*

She ends the interview with a very strong statement about her intent to take control of her learning:

> *But I also have to make the time to figure out what I need to do and get someone to teach me. I don't think it's difficult; it's just something else to learn: a whole other set of techniques.*

Shortly after this interview, A22 attended a major tissue engineering conference. This opportunity was the turning point for her; she decided to apply and transfer to the PhD program.

Postscript: Gender-Focused Interview 12–11–2003

Approximately 9 months after the previous interview, as part of the supplementary project on gender experiences, we conducted an interview with A22 that focused more directly on her experience of gender-related issues in Lab A. As noted in Chapter 6 on race and gender enactments, the inescapability of gender means that every person's history is in part gendered. This includes the person's learning history; thus we cannot separate gender from the enactments that constitute learning. We conclude our discussion of the learning person with a postscript that includes highlights of this interview. In keeping with our analysis, the postscript is not intended to "wall off" gender, only to indicate that it reflects a different kind of interview from those we have been quoting from thus far – one meant to explicitly elicit her thoughts on gender in relation to her work.

The interview occurred a few days after the interviewer had presented her preliminary analysis of gender-related discourse in Lab A in a public lecture at which A22 was present (see Chapter 6). The analysis in the presentation was based on recording and analysis of field observations of casual conversations in Lab A. The interviewer begins by asking A22 for her thoughts on and responses to the talk. A22 replies that the talk "*was interesting on several levels.*" She does not elaborate on what these levels are or on the different ways in which the talk was interesting to her, but says,

> I must say it was kind of hard to get used to hearing things that you knew you, that knew came from your lab, and you had to kind of get past that, and it was interesting seeing how things are interpreted.

This is followed by what appears be something of a denial of the interviewer's interpretations:

> You only see part of it. Because you don't necessarily see how we interact with each other all the time. And then you take that, and then you take parts of that, and show it to other people – and it's just interesting how, how that comes off sometimes . . .

The interviewer points out that, although her analysis is preliminary, she is interested in ways that gender affects relationships and experiences in the laboratory. She explains that she is interested in more than just discrimination, but rather, how "being a woman" is manifested in laboratory practices in ways that are not always obvious. The incident A22 refers to in the following passage is related to the comment, "*you sound like a waitress,*" made by one female student to another in a lab meeting that was discussed in the section on gender in Chapter 6. In comparison to her confident responses to the previous interviewers when discussing her research and technique learning, A22's responses to this interviewer are rather halting and meandering:

> *The one that I think struck me the most was the conversation that was brought up, that was obviously A26, umm, [laugh] becau – and I know that because one of those ca –, parts of it, she was discussing a conversation that we had had – and I know, I was like, I don't even know if you knew that or not, and, and I talked to A26 later, and I don't even think that she remembered that it was me, but the rea – maybe it was with other people as well, I just know that it's a conversation. . . . I just know that it's also a conversation that I know she and I have had as well as myself and other women, and some of men, that I have worked with have had off and on in my career, as well. . . . As well as women, uh . . . working in the workplace, and her –, some of her comments about flirting versus non-flirting – [I: Right.] – And, and I know that's something that, that I noticed – and, and, I li-, that was not, I was like, hmm and, you know, because A26 and I have, we have different personalities and we treat things differently. . . .*

A22 then recalls an observation at one of her previous jobs in which she was supervised by a woman who impressed her as "*really trying to be one of the boys.*" She tells the interviewer that she is not really sure what she means by that, but that it is an episode she was thinking about after the interviewer's talk. She elaborates on this incident, questioning what it was that made her feel that way about the other woman. She decides that it has something to do with the hours the woman kept and "*some of the things she tried to do,*" including "*going to lunch at somewhere I wouldn't go to lunch.*" She tells the interviewer that she was "*consciously saying to myself that's not going to work for me. I can't do that . . .*"

A22 tells the interviewer that at present she feels "*so new*" in the lab [90% female] and that she is "*trying to get up to the same level as everybody else,*" but that she has made "*conscious decisions*" throughout her career "*on how to behave around groups of men.*" She notes that her husband would call her style of interaction "*flirtatious*" but that she herself would not, adding that "*some of it is just who I am and how I interact – but some of it is conscious – in order to get my job done.*" She relates some personal history to the interviewer, describing a job setting in which she was the only woman. She recalls in one setting there was a technician with whom all the engineers found it difficult to work: "*You had to be nice to him, because if he was mad at you, your work wouldn't get done.*" She adds,

> *Being a woman, how I interacted with him was, was very different.... I'd watch guys go down there and be rude, and they're like "Well, this VP says that I need to get this done," and they would get things done, and I couldn't get away with that at all! But if I was nice, I'd get it done – I got to the point where I was treated differently, for whatever reasons, and, if that worked for me – I was going to take advantage of it.*

A22 expresses to the interviewer her inability to sort out how much of the different treatment she received was due to her sex and how much to "*personality.*" She notes that she and her husband had very similar jobs before they were married, but people treated them very differently. She is not sure to what extent this different treatment reflects personality or "*style.*" She notes that her husband comes across as very self-assured and even "*authoritarian,*" traits she does not see herself as having. She adds that "*statistically men are more likely to act like they know something when they don't.*" She continues to puzzle over the extent to which the differences she notices are due to gender or personality:

> *But at the same time my way, my method worked very well. I mean, I had VPs who knew who I was and would say hello to me in the halls, because I would just be – I think it's the whole fostering communication and, and, and, umm, relationships. Because a lot of the flirtatious nature is really just being nice to people and getting to know them and so ...*

The interviewer then changes direction, inquiring about A22's reaction to the fact that she is now working for the first time in an environment that has a dominant number of women. A22 tells the interviewer that the environment is different for her, not only because of the larger number of women but also because for the first time she does not feel in "*gender competition,*" meaning that she does not feel the same need to "*prove to yourself that you could be as good as the guys.*" In fact, she says that it is interesting to listen to the one male graduate student researcher in Lab A talk about his experience, because "*he must sound like some of the women*" in A22's previous industry (mechanical engineering). Here she introduces a slight note of uncertainty, saying that "*it's really interesting, because sometimes he's kidding – and sometimes he's not. And I'm not sure I know when – when....*" She calls this "*almost a role reversal*" and then expresses concern that the male researcher might be feeling excluded.

The discussion then turns to broader patterns of social interaction among Lab A researchers, including which ones tend to go out for lunch together and the factors that affect those patterns, such as income differential or marital status. A22 then invokes the physical space of the laboratory as contributing in an important way to the communication patterns:

> *A lot of it's just because where people sit. Where, where I am is where a lot of the shop talk and not shop talk goes on, because we've got A5 and A7, and they're both so far along, and A26, you know, and because they are both, are senior people in the lab who are highly respected, a lot of people will come there and ask them questions. A7 also has her, uh, has* [her mentee] *who comes back and asks question and may, you know – so, a lot of times* [A26] *will be back there and I'll be talking about something that* [A26] *has a question about, or something that* [A5] *has a question about, and wants feedback from other people, and it's, it's a very spontaneous – there's a little bit of that that goes on in the other areas.* [All researchers mentioned are female graduate students.]

The interviewer then comments that researchers in Lab A generally seem happy and that there is a high retention rate in the lab. A22 says that she thinks "*a lot of it is the communication, because you ha-, you can really, like, if I need something I know I can ask anyone in the lab at any time.*"

When the interviewer then asks about any challenges A22 faces, she mentions the difficulty of balancing class work with research ("*school versus lab*") and also that she feels a need for more guidance, perhaps from one specific mentor or the lab director. She recasts the difficulty into an advantage, however:

> *And as much as it's very difficult and frustrating right now I know I'm gonna be better for it – I'm gonna be much better off than some of my other fellow grad students who didn't have to choose their projects, who had them given to them and told here study this, because I'm going to have to decide – I'm gonna have to write a grant and get money and come up with an original idea on my own – I already have several different areas I could go to, you know, it it's helpful because I've already had to do that to some level.*

Later on, the interviewer asks A22 to comment on "the other aspects of life." A22 responds that she thinks that "*having a life outside of your work is very important.*" She continues by saying, "*I want to be more than just my job I've always wanted to have fun. I want to have a family. There is so much more to life. If I could get paid to hike all the time I might consider doing it.*" However, she acknowledges, "*It's a lot harder now because I know that in order to graduate I'm going to need to dedicate myself.*" She emphasizes that she hopes to be passionately engaged with her work, despite her desire to want more than just her job in her life:

> *After watching my husband who's got his passion and watched some of the other people I know who really knew what they were passionate about, I realized that I really do want to do something where I am passionate about my work. But I don't want it to be the only thing; I don't want it to define me.*

She says that, unlike her husband, she's not sure she's found her "*exact passion*" and calls this frustrating. However, she mentions that an opportunity is arising to work with a pediatric surgeon and that the burst pressure testing applications she is working on have the potential to be of great benefit through this collaboration. The idea would be to find a way to develop artificial tissue that could be implanted in children, tissue that could "*grow with*" the child instead of requiring regular replacements. She said that she thinks the use of artificial tissue could help reduce the need for children to undergo surgery every 2 years. Her expression of emotion is strong in discussing its potential:

> *I am passionate about it because* [it offers] *the chance to actually make a difference – some of the work that we do could potentially help scientists in other fields and to know that I am doing something that can help other people and potentially make the world better then it's something I can be passionate about. And can make me feel good about the work I'm doing and I – it feels important and I'm interested in it and I like talking about it.*

It would be impossible to attempt an analysis of the aspects of A22's learning trajectory and value hierarchy that specifically reflect gender. Yet, clearly, gendered relations have marked her experience, and in some respects she has conscious access to these markings. Being a woman affected her mode of interaction with a supervisor in industry, particularly because she was the only woman in that laboratory. Her husband has labeled her way of communicating as flirtatious, but she sees her actions as reflecting merely a relationship-fostering agenda. She even names the experience of needing to prove herself to be competent as gender competition. However, in keeping with our earlier remarks on gender enactments, A22 also reflects on the particular ways in which she has observed other women in science dealing with what is at least in part a matter of gender. This is summarized most cogently in her remark that she and A26 treat things differently because of their different personalities, despite their acknowledgment of commonalities in their experience as women in an environment traditionally staffed by men.

SUMMARY AND CONCLUDING REMARKS

We chose to conclude the main portion of this book with a case study of learning because, although A22's experience provides only an illustration, it is an important one for the major point of this book. The detailed examination of her interviews in sequence illustrates precisely the need for *integrated frameworks* in science studies and psychology. A22 is before all else an acting person, uniquely constituted, privately experiencing, and shaped by a particular history. At the same time, she acts within the social and material constraints of her laboratory practice and a set of conceptual and social categories with histories of their own: "scientist," "mechanical engineer," "woman," "wife," even "older graduate student."

The series of interviews with a single researcher in Lab A underscores first of all that research laboratories are informative settings for investigating situated learning. The complex interdisciplinary structure and innovation focus of the laboratories under investigation provide particularly rich environments for the study of learning "on the ground." Moreover, careful examination of the interviews reveals that the study of learning in these research laboratories requires attunement to social, cognitive, affective, motivational, and material dimensions of practice. Learning is at once fully "personal," inasmuch as her passion, engagement, commitment, frustration, and desire for belonging and meaningful work are privately felt and communicated only imperfectly to the interviewer. It is to *this* researcher that her unique history belongs, her hands that are performing the tasks through which she sees her learning as progressing, her brain that is registering changes. At the same time, the interviews demonstrate how thoroughly A22's research is situated in the social and material configurations of the particular laboratory in which she practices tissue engineering, in American laboratory science in the early part of the 21st century. It is emergent in the formal and informal mentoring relationships that enable the researcher to observe, ask questions, seek approval, obtain emotional support, and feel a part of something important. It is transactional in that she also provides consultation to others and nurtures the cells that are the foundation of her research and learning. In turn, working collaborations and the cells themselves are crucial to her aspirations to use tissue engineering to help children, thereby sustaining her challenging work and study schedule with a felt sense of purpose.

In addition to underscoring the need for integrated frameworks, several more specific observations can be made from the analysis of A22's interview data pertaining to learning in a research laboratory that have bearing on the study of learning more generally:

1. Laboratory learning is constituted through doing, through performance of tasks both novel and routine that make discoveries and innovative applications possible.
2. Laboratory learning is facilitated by a social environment in which the researcher can observe, ask questions, obtain supervision and feedback, and experience a sense of community.
3. Laboratory learning accompanies transformations in identity and the experience of competency, ownership, and intention for which it is conventional to apply the term "agency."
4. Laboratory learning is affectively punctuated and sustained.

In sum, effective learning environments are those that enable learners to find a niche, to feel a sense of ownership and control, to experience a sense of belonging, and to establish informal relationships that facilitate questioning and observing and have a positive impact on the affective dimension of learning. We base this claim on the fact that in both laboratories learners were successful along all the standard measures for scientists – conference presentations and posters, publications, qualifying examinations, and dissertations – and all currently have positions in academia or industry.

A more complete integrated account would provide, in addition, more detailed analysis of ways in which the physical layout of the laboratory and the material properties of the objects and devices used for investigatory purposes contribute to learning and research – topics that are part of our ongoing research and that require additional data collection and analysis methods. We do not claim that interview data alone are sufficient to provide the kind of integrated account we see as not only desirable but also necessary for understanding "laboratory life." Yet without the personal accounts of learners, we clearly lack any possibility of full integration. We contend that our analysis demonstrates that important point effectively and in so doing puts flesh on the venerable wisdom of Dewey (1938a) and of the generations of educational and philosophical commentary on his insights – emphasizing that knowing and doing are functionally mutual, that learning is tied to both bodily manipulations and social interactions, that learning and emoting are intertwined, and finally that "everything depends on the quality of the experience that is had" (Dewey, 1938a, p. 27).

8

Epilogue: Science as Psychology: A Tacit Tradition and Its Implications

In this work we have explored the complex interrelations of several categories of human activity that continue to occupy different domains of scholarly discourse in both science studies and psychology. The problem with the conventional separation of activity categories – cognition from emotion and emotion from identity and identity from learning – is that it tempts one to assume that the variously described activities function independently in practice. In our analysis of interview data from two interdisciplinary biomedical engineering labs on the campus of a leading research university, we have affirmed the acting person as our analytic unit, while fully affirming that all activity occurs within contexts that are cognitively, socially, and materially normative. A laboratory is one such context, with methods, procedures, and protocol embedded within wider contexts, professional and historical.

Although we consider interview material to reflect practices grounded in and constrained by the interview event rather than mirroring images of the enduring inner worlds of our participants, we are nevertheless obliged to take researchers' accounts of their science practices to be important forms of science practice in themselves. Thus we can make the following claim quite boldly, and we invite consideration of its important implications for the science we study and for the unit of analysis we engage: *Glaringly evident in the interview material is that cognitive, social, material, cultural/historical, and affective dimensions of human practice intertwine in ways that are vital to problem solving and innovation in the laboratory.* This statement is not intended to compromise the integrity of the research process or to call into question the epistemic authority of what is produced by this intertwining activity. It is merely to assert that the contextually embedded acting person is the engine and oracle of science.

In so affirming science as an activity of persons, and in viewing activity as an inherently integrating category, we align our work with many earlier perspectives on science and psychology. In this final chapter we situate our efforts within a largely unappreciated tradition of person-centered accounts of science and an implicit turn to science as psychology along the lines here construed. To do so, we need to also situate it within the newly forming field of psychology of science.

PSYCHOLOGY OF SCIENCE AND SCIENCE AS PSYCHOLOGY

Broadly understood, this volume constitutes a contribution to the field of psychology of science. However, our analysis differs methodologically from standard psychological methods, which typically involve isolation of variables to discern and predict their interactions. Instead, a fundamental assumption contributing to the conceptual and methodological framework engaged in the book is that one of the best forms of data available to psychologists is that which cannot be controlled – that which is obtained by talking to and observing persons engaged in "real-world" practices. Our contribution to the psychology of science therefore reflects efforts to analyze interview data collected as part of a multiyear investigation of biomedical engineering laboratories on the campus of a research university for which the second and fourth authors are principal investigators. In their efforts to describe the cognitive and learning practices of scientists at varying levels of expertise within these laboratories, Nersessian and Newstetter have already made substantial contributions to cognitive science, the philosophy of science, the learning sciences, and science education.

Our focus and intent in the present book have been to diverge somewhat both from those efforts of the larger project and from current research in what is emerging as the psychology of science. Our analysis is not geared toward understanding what is special and unique about scientific cognition or culture so as to make possible the special form of knowledge that is science. Rather, our grounding assumption is that science *as human practice* is an undertapped resource for generating new understandings of human psychology in general, that the analysis of science can inform questions concerning the practices of persons in general as much as it tells us about scientists themselves.

Effectively, we are offering what amounts to a shift in the means by which the relation of psychology to science might be construed. We are interested in how the study of scientists in practice might help us better understand human beings and to develop psychological theory that transcends the boundaries of what we traditionally identify as science. The book's focus on science *as* psychology necessitates a deliberate evasion of the question of whether psychology can or should be a science. We have turned instead to the question of how studying scientists in practice can yield insights that are importantly psychological. The significance of this question is grounded in a view of science as a normatively structured form of community, within which intentional persons act and make sense of their actions for themselves and others. Although this is an elite and highly specialized community, its practices – namely, cognitive, investigative, and interactional practices leading to the production of knowledge – are exemplary of what it means to "adapt" to the world, to change it, and to be changed by it. Following this way of thinking about science, we consider what "real-time" science practices can tell us about the fortunes and complications of being human.

In offering this analysis of science practice we latch onto an underappreciated, unrecognized tradition by which scientists are used as a model for understanding human beings more generally (science as psychology).

Science as Prototypically "Human"

In arguing for the merits of psychology of science as a research focus, Greg Feist highlights the status of science and scientific thinking as "prototypes of human thought and understanding" (Feist, 2006a, p. 3). Although we concur wholeheartedly on the importance of the new specialty and the prototypical status of scientific cognition, we note that the implications of the view of science as prototypical are worth examining carefully.

Of course, discussion of prototypes is historically bound to the topic of categorization; that is, the basis on which things in our experience are grouped together as a set. This topic is of such fundamental importance to the greater history of Western philosophy that we could not possibly begin to treat it responsibly here. Moreover, prototype theory is a major focus of historical and contemporary cognitive science. To be brief, even in common parlance "prototype" has at least two quite distinct connotations: one implying an idealized form of something, the other a typical representative of a class. Two quite varied accounts of scientific thinking correspond to these two connotations. In turn, the different accounts of scientific cognition give rise to two different depictions of scientists and the human subjects scientists represent.

Prototype as Ideal

A "prototype" may be understood as an idealized model, in the sense that Plato's forms constitute ideal and enduring prototypes in comparison to which particular instantiations of the kind of thing in question fall short. No particular instance (an actual cat) ever perfectly resembles the ideal form (the prototypical cat). Yet the quality of any particular instance of the thing in question might be evaluated in relation to the ideal. The form of virtue thus determines the extent to which any act might be considered virtuous, and so on. This sense of prototype is similarly related philosophically to essential qualities, prompting discussion of categorization in terms of necessary and sufficient conditions for inclusion in a class (after Aristotle, *Categories*).

In varying degrees of explicitness, science has been regarded as the superlative human achievement for much of psychology's controversial history, setting the standard for scholarly accomplishment and providing the "form" against which other modes of cognitive operation might be evaluated for bias, trickery, and error. Ideas about science have long been entangled with assumptions about access to reality, the possibility of overcoming individual or collective interests, attaining objectivity, and faith in the power of human agents to *make progress*. Progress is imagined in the possibility of distilling rationality to a pure form able to transcend cultural and historical situations or the particularities of personhood (subjectivity). Philosophical reconstructions of the essential inner logic of science vary in specific details and emphasis, but their points in common are neatly summarized by Margolis (2003) in a description of "scientism" in 20th-century philosophy of science:

> "Scientism" signifies the assured position of a privileged methodology or mode of perception, or even the assured validity of a metaphysics deemed ineluctable or overwhelmingly favored by the self-appointed champions of "Science" even in the face of insufficient evidence or substantive doubt or their proclaimed opposition to cognitive privilege or Cartesian foundations (p. 6).

Scores of critics have challenged the accuracy of the received view of science; some have depicted it as compelling and powerful fiction. Ian Mitroff (1974), for example, derides what he calls the "storybook image of science," demonstrating the proliferation of this image with a series of historical examples from science textbooks. Its most prominent features are emotional neutrality and detachment, occasionally to the point of surrendering or "renouncing" emotion and personal interest to the cause. Richard

Lampkin's 1938 article entitled "Scientific Attitudes" is particularly illustrative: "The scientist deliberately renounces all emotion and desire, except that of accomplishing his twofold purpose. The scientist should never make pontifical announcements, nor indulge in melodrama. If the scientist is ignorant on certain points, he must acknowledge his ignorance on those points" (quoted in Mitroff, 1974). Mitroff's review of relevant literature reveals Lampkin's appeal to a disinterested, dispassionate ideal to be widespread in sources used to educate budding scientists. In one case there is even a call for scientists to control their imaginations and resist the lure of novelty! The naïveté and simplicity of this view of science, widespread though it might be, are what Mitroff intends by the "storybook" label:

> It is a Storybook view because it is the kind of picture that we should expect of children; it is an image of science that is lacking in sophistication or any deep insight into the nature of science. As incredibly naïve as this image of science is, it is important to emphasize that I am not talking about or concocting a straw man.... It would be easier to dismiss the Storybook view outright if it had not been held by so many intelligent and otherwise discerning men of importance and influence (Mitroff, 1974, p. 8).

Similarly, the broad cultural and academic glorification of science is captured by Philip Kitcher's appeal to "Legend:"

> Legend celebrated science. Depicting the sciences as directed at noble goals, it maintained that those goals have been ever more successfully realized. For explanations of the successes, we need look no further than the exemplary intellectual and moral qualities of the heroes of Legend, the great contributors to the great advances. Legend celebrated scientists, as well as science (1993, p. 3).

Scientists celebrated in and by "Legend" display an uncompromising commitment to truth, possess reliable cognitive abilities, and execute strategies of knowledge acquisition that render their discoveries and applications value-free and immune from social contamination. For Michael Mahoney,

> The scientist is the high priest of knowledge. He is our capable and mysterious liaison with reality, our ambassador to wisdom. Moreover, the stereotyped image of the scientist portrays him as a superhuman organism. He is viewed as the paragon of reason and objectivity, an impartial genius whose visionary insights are matched only by his quiet humility. He combines the wisdom of a prophet with the dedication of a martyr (2004, p. 3).

Much has been made of psychology's struggle for standing as a natural science and the yoking of its very identity as a "modern" discipline to the idealization of science and the promises science holds for human advancement. Knorr Cetina (1999) suggests that pivotal to the idealization of science is the fantasy of unification *in* science; the idea ("myth") of unification continues to inform the social sciences and their position in relation to this imagined entity. That the ideal upheld, not to mention the very problems and methods undertaken as consonant with this ideal, is very much a product of the 19th century is widely acknowledged by historians.[1] Similarly, the methods psychology envisions as comprising good science reflect an effort to pattern after the ideal. Harré (1998) trenchantly observes that many psychologists engage "not the real methods of natural science but some imitation of their superficial features" (p. 30).

Nevertheless, an idealized view of natural science and scientific rigor has inspired psychology's canonized methodological commitments:

> The methods of science have been enormously successful wherever they have been tried. Let us then apply them to human affairs. We need to retreat in those sectors where science has already advanced. It is necessary only to bring our understanding of human nature up to the same point. Indeed, this may well be our only hope (Skinner, 1938, p. 5).

However, the influence of science as a normative template for understanding all things psychological is even more pervasive. The ideals of science – more specifically, the powerful mythologies evinced by science – inflect the very nature of how psychologists think of cognition, including social cognition, and the directions and disputes of cognitive inquiry. Heuristics – shortcuts in cognition that guide the average soul in making predictions, attributions, and decisions in relation to social interactions – are defined most frequently by their deviation from scientific thought. In their extensive review, Kruglanski and Ajzen (1983) note that "the scientific method in general [is viewed] as superior as or closer to normative than lay modes of inference and [it is] suggested that one way in which the tendency to err might be mitigated is instruction of lay persons in the canon of scientific (e.g., experimental) methodology" (p. 3). Thus, for example, judgments about uncertainty are evaluated and framed in terms of Bayesian statistics; everyday judgments described as "irrational" are those for which the influence of probability, sample size, and other systematic considerations is underrepresented (Tversky & Kahneman, 1974, p. 1125). The legitimacy

---

[1] See Danziger (1990); Koch (1992); and Robinson (1989).

of taking an assumed view of scientific rationality as the means by which cognition is evaluated goes unquestioned.

Psychology's tenacious clutching onto an imagined science remains tied to the conception of scientific cognition as prototypical in the sense of highest or ideal, with its strivings toward overcoming bias and attaining objectivity. It is a profound understatement to note that challenges to the very possibility of this attainment are many, as are doubts about the desirability (ideal status) of the very sterility incumbent on such a view of science. The science envisioned in what Kitcher has called those "dear dead days" (1993, p. 3) has engendered labels ranging from "ethnocentric" to "patriarchal," "naïve" to "violent." Assaults on the assumption of the superiority of science have bled into rejection of the ideals of truth and knowledge as desirable endpoints, and "progress" as the appropriate depiction of scientific and technological advance (Feyerabend, 1975). Taken to an extreme, science is cast as a threat, an arid but pernicious force in oppositional relation to the humanities, holistic psychologies, and all that is appropriately messy and irreducible. Although late-20th-century scholarship in philosophy and cultural studies is most readily associated with the attack on science as mechanistic and dehumanizing, we must recognize this attack as a resurfacing of a sentiment that virtually accompanied the original successes of "modern" science and has never fully dissipated.[2] The idea of science as liberator or ravager of humanity similarly issues from a conception of scientific cognition as prototypical in the sense of idealized and abstracted. The cultural depictions of *scientist* that follow from this idealized conception of science range from the visionary hero to the mad destroyer to the "nerd" with narrow interests and constricted soul. Thus, according to Mahoney, although science remains a prestigious profession carrying substantial social status, "the popular stereotype of the scientist is reminiscent of character portrayals in children's storybooks" (2004, p. 3).

Prototype as Typical

Alternatively, a prototype and that which is prototypical might be thought of as illustrating the *typical* qualities of a class (according to *Dictionary.com Unabridged*). "Typical" is as far in meaning from "ideal" as is possible. Of

---

[2] Thus Toulmin referred to the 1960s and '70s as offering "the renewal of an attack on the mechanistic 'inhumanity' of Newtonian Science launched 150 years earlier by William Blake in England, and Friedrich Schiller in Germany" (Toulmin, 1990, p. 7). Clearly Vico, Rousseau, Dickens, Dostoyevsky, and legions of others might be counted among the antagonists of scientism and its broader societal effects.

course, a case might be made for thinking of both senses of prototype as stemming from the notion of abstraction from particular examples of a class, as is the case when the prototype of dog is constructed from numerous personal experiences with four-legged barking creatures. Well-known experiments by Rosch (1975) suggest that categorization is a graded phenomenon, with some members of any class regarded as more representative than others and the most typical understood as prototypical. For instance, a robin is a more typical representative of the category "bird" than a penguin. Thus, this is a much different notion of prototype.

How does shifting the sense of prototype to "typical representative" alter the conception of science as prototypical and the implications that follow? Science as prototypical in the second sense suggests a domain of activity that is ordinary and everyday, rather than that which constitutes the farthest reaches of human possibility. As an everyday or typical example of a human enterprise, we would expect science to be fraught with everyday passions and aspirations and mishaps and serendipity; to be influenced by social, economic, and political structures; and to be guided to some degree by tacit perceptions and conflicting energies below the surface of awareness.

A competing representation of scientific cognition and of the process of science in keeping with this second sense of prototype as typical or ordinary is available, if only comparatively recently so. Writing in the mid-1960s, Storer identifies the turn to the idea that "scientists are just like the rest of us" as of recent origin (Storer, 1966). It is also an idea pivotal to the efforts to understand science as fundamentally a social rule-following, economically driven system. Philosophically, sociological accounts were bolstered and extended at mid-century by Kuhn (1962) and later by the "strong programme" of epistemology (Bloor, 1976) and other well-known movements. A more recent shift in focus to the "construction machineries" that constitute science has extended the application of the everyday sense of science as prototypical (Knorr Cetina, 1999, p. 11).

ENGAGED SCIENCE: POLANYI AND MASLOW

The idea that science is prototypical in the sense of an ordinary representative often is, but need not be, coupled with rejection of the special epistemic status of scientific knowledge. For example, Polanyi's revisionist account of science process offered in *Personal Knowledge, Science, Faith and Society* and *The Tacit Dimension* never denies the existence of natural laws, but challenges the received account of the means by which they are apprehended. In his preface to *Personal Knowledge*, Polanyi explains,

I start by rejecting the ideal of scientific detachment. In the exact sciences, that false ideal is perhaps harmless, for it is in fact disregarded there by scientists. But we shall see that it exercises a destructive influence in biology, psychology, and sociology, and falsifies our whole outlook far beyond the domain of science. I want to establish an alternative ideal of knowledge, quite generally (1958/1973, p. vii, emphasis added).

Polanyi's "conceptual reform" of science is rooted in Gestalt psychology, which he views as "an active comprehension of things known, an action that requires skill" (Polanyi, 1958/1973, p. vii). Skillful performance (equated with skillful "knowing and doing") entails "subordinating a set of particulars, as clues or tools, to the shaping of a skilful achievement, whether practical or theoretical. They are made to function as extensions of our bodily equipment and this involves a certain change of our own being" (p. vii). Elsewhere he calls discovery "an extremely delicate and personal art which can be but little assisted by any formulated percepts." (1946/1964, p. 34).

Polanyi is adamant that his reframing of knowledge as personal participation does not equate to an account of scientific knowledge (understanding) as *subjective*:

Comprehension is neither an arbitrary act nor a passive experience, but a responsible act claiming universal validity. Such knowing is indeed *objective* in the sense of establishing contact with a hidden reality; a contact that is defined as the condition for anticipating an indeterminate range of yet unknown (and perhaps yet inconceivable) true implications (1958/1973, p. viii).

Polanyi regards this knowing as a "fusion" of the personal and the objective, to which he attaches the label "Personal Knowledge." The book that follows from this premise offers reformed conceptualizations of objectivity, probability, order, articulation, and affirmation as scientific values by tying them to passion, commitment, skill, value, and conviviality in the production of scientific knowledge. Idiosyncrasy is tied to epistemology, particularly through the imparting of *value* to the objects of inquiry. Through passion, through interests – that frequently cannot be defended on any rational grounds other than that they *are* interesting to the scientist – some objects, aspects of objects, and problems are selected out as meaningful and worthy of investigation, and some results are signaled as important. However, of course, we might note that passions are influenced by communities; that is, by a scientist's particular history of participation within a series of communities (graduate school, research laboratories, industry settings) that affect what he or she *feels* to be a worthy or important problem (Machamer & Osbeck, 2004).

Polanyi's emphasis on the creative, personal, and intuitive dimensions of science provided great inspiration to Abraham Maslow, as shown in the interesting essay published as *The Psychology of Science* in 1966. This work, albeit critiquing a view of science as impersonal and mechanistic, similarly warns of the dangers of unbridled immersion in experience (the shadow side of humanistic psychology's search for a more meaningful psychology). Maslow sees his project as "trying to enlarge science rather than to destroy it" (1966, p. xvi), with the overarching goal being one of trying to "rehumanize science" (p. 3). Among the targets of critique in Maslow's elevation of "interpersonal (I-Thou) knowledge as a paradigm for science" (p. 102) is the notion of science as a value-free endeavor. This he relates to the ideal of scientific detachment as cool neutrality, which he regards as contrary to his own ideal of science as experientially engaged encounters with the objects of inquiry. The polemic Maslow constructs here is between objectivity and concern; he reads detachment as disengagement, even apathy in the service of neutrality of stance. However, Maslow acknowledges the necessity of this emphasis on detachment as it emerged historically within a context in which the real interference of doctrine and tradition posed not only a threat to knowledge but also to persons: Bruno and Galileo being the most obvious examples.

In the end Maslow promotes what he regards as an integrated version of science, whether of natural or social objects, whereby one follows rules but remains passionate, involved, creative, good-humored, humble, and skeptical of one's own accomplishments, even as one is transformed by the objects and processes with which one fully and deeply engages. In this way Maslow offers us a kind of Polanyi-lite, one whose principal target appears to be fellow psychologists whom he accuses of smug "methodolatry" (p. 145). Important to the point at hand, in his effort to rehumanize science Maslow offers an account of science as prototypical in the second sense identified:

> One trouble with defining science in terms of its highest reaches and ulti-mate skills is that it makes science and the scientific spirit inaccessible to most people. . . . It comes to be seen as a matter for the expert, something done by a certain kind of highly trained professional and by nobody else. . . . If we define science in terms of its beginnings and its simplest levels rather than in terms of its highest and most complex levels, then science is simply looking at things for yourself rather than trusting to the a priori or to authority of any kind. . . . It follows that a child can

be "scientific" watching an anthill and so can a housewife comparing the virtues of various soaps by trying them out in her basement.... And when phrased in this way – keeping in touch with reality, keeping your eyes open – it becomes almost a defining characteristic of humanness itself (pp. 135–136).

PERSON AS SCIENTIST[3]: GEORGE KELLY

The depiction of the empirical attitude as closely associated with or even typifying human nature is also strongly suggestive of George Kelly's *man-as-scientist* metaphor as developed in *The Psychology of Personal Constructs* (1955). It is the "everyday" conception of science that George Kelly imagines in establishing scientific thinking as his core metaphor for human functioning. At first glance the metaphor seems to reflect a sterile, detached, purely rational conception of personality, and this is frequently the reaction of students on hearing it for the first time. Yet a closer examination of Kelly reveals a rich and multifaceted understanding of science and scientists, one in which caring deeply, engaging passionately, and committing fully and personally to the objects of inquiry form no contrast class with systematic procedure. This vision of science in turn led Kelly to reconceptualize basic psychological categories as they relate to persons and scientists.

The conceptual grounds on which we conceive of "science as psychology," as offering contributions both to science studies and general psychology, are largely informed by the underappreciated implications of Kelly's person-as-scientist metaphor. It is thus important to the overall project of this book to examine carefully what Kelly appeared to intend by this metaphor, as well as its overall resonance with the claims we have been forwarding through our analysis of practicing scientists.

Overview: Kelly embraced a notion of science as prototypical in the sense of being closely aligned with ordinary and everyday level human ways of being:

> When we speak of *man-the-scientist* we are speaking of all mankind and not merely a particular class of men who have publicly attained the stature of "scientists."... We are speaking of aspects of mankind

---

[3] Here there are similarities to the "child as scientist" research in cognitive development, wherein young children are characterized as "intuitive scientists" gathering knowledge about the physical and social world. See, for example, Carey (1985); Gopnik (1999); and Wellman (1990). This research is outside the scope of our discussion, but we want to note the relevance of its framework and assumptions to the broader theme of "person as scientist."

rather than collections of men. Thus the notion of man-the-scientist is a particular abstraction of all mankind and not a concrete classification of particular men (Kelly, 1955, p. 4).

Kelly's interest in science fueled his theory of personality and general psychological functioning, one that he likewise developed into an innovative approach to assessment and psychotherapy. Kelly's exploration of the commonalities of scientific practice and everyday human practice has been credited to his own early immersion in physics and mathematics as he originally prepared for a career in engineering (Fransella, 1995). As Kelly moved through a series of career turns to his eventual fundamental interest in the human sciences of personality and psychotherapy, he retained his overarching interest in science and revised notions of objectivity rooted in quantum theory and relativity that were still considered revolutionary in his time. Kelly places the sense-making activities of the scientist at the center of his theory of personal constructs, as a model for the ways in which constructions of reality are created, tested, and revised.

But what does Kelly mean by a "scientist" precisely, and why does this conception of science afford new understandings of human functioning more generally? An initial reading of Kelly's person-as-scientist metaphor foregrounds *anticipation* and *control* in human motivation and personality, central as these are to all science. Lay "theories" are constructed on the basis of experience and are translated into hypotheses that are tested against evidence and revised as needed. The ultimate aim is prediction and control, and it is with these features that Kelly begins his description of the scientist:

> Let us see what it would mean to construe man in his scientist-like aspect. What is it that is supposed to characterize the motivation of the scientist? It is customary to say that the scientist's ultimate aim is to predict and control. This is a summary statement that psychologists frequently like to quote in characterizing their own aspirations. Yet, curiously enough, psychologists rarely credit the human subjects in their experiments with having similar aspirations (1963, p. 5).

Thus prediction and control are the standard elements of personal construct theory's person-as-scientist metaphor as distilled into introductory textbooks on personality.

The person-as-scientist metaphor can be best understood as a means of understanding the essential sense-making orientation of persons, an orientation that necessitates an active pursuit of tentative working order in a maelstrom of chaos and uncertainty. The evolving, active, practice orientation of Kelly's theory aligns it with earlier pragmatist conceptions of human knowing; both Kelly himself and his biographers noted the influence of Dewey in particular (e.g., Fransella, 1995). Indeed, elements of the person-as-scientist metaphor might be identified in Dewey's own conception of knowledge and activity as inextricably fused and in his emphasis on the exploratory, problem-resolution focused features of thinking and reasoning (Dewey, 1910). For Kelly, the central human activity is one of *construing* – a continuous, committed, engaged interpreting process that mobilizes the whole being across every domain of life.

The central units of analysis in personal construct theory (PCT) are the *construct* and the systems of constructs that collectively function to structure experience: "Each individual man formulates in his own way constructs through which he views the world of events. As a scientist, man seeks to predict, and thus control, the course of events. It follows, then, that the constructs which he formulates are intended to aid him in his predictive efforts" (Kelly, 1963, p. 12). Much has been written by Kelly and others on what exactly he meant by a construct and how constructs differ from other units of analysis such as representations, concepts, and symbols. According to Kelly, a construct is a "representation of the universe," but one that is created rather than imposed by contact with this universe (i.e., a copy). A construct is "a representation erected by a living creature and then tested against the reality of that universe. Since the universe is essentially a course of events, the testing of a construct is a testing against subsequent events . . . in terms of its predictive efficiency" (p. 12). The universe comprises both social events (interactions) and material events, though there is an implicit sense in Kelly's theory that, as for G. H. Mead, the physical universe within which one acts is importantly "social." Constructs are "reference axes, upon which one may project events in an effort to *make some sense* out of what is going on" (unpublished manuscript quoted in Bannister & Mair, 1968, p. 16, emphasis added). In this same passage Kelly calls a construct "a cleavage" to emphasize its function of making discriminations in experience.

Because cleavages, discriminations, and "channelings" of the kind Kelly proposes occur at both the preverbal and nonverbal levels, construing is not purely a propositional activity. Rather, it engages and is inseparable from the primary acts of perceiving. Neither is constructing an activity separated or abstracted from the affective realm. Fransella (1995) summarizes the connection of construing to perceiving as follows:

> Constructing is not "thinking" or "feeling" – it is the act of discriminating experientially: It is the way we have perceived – at some level of awareness – that certain events around us are repeating themselves and

are thereby different from other events. Once we have noticed similarities and differences (discriminated) between events, we can anticipate future events. . . . Thus, experiencing and construing are part and parcel of the same process (p. 53).

The view of constructing as a form of making discriminations in experience forges important links between personal construct theory and the ecological approach to perception (Gibson, 1979) that has been a formative influence on the distributed (and distributing) cognitive frameworks discussed in the book's introduction.

The psychology of personal constructs built around the person-as-scientist metaphor and the assumption that construing is a primary human activity are the basis for a detailed and systematic theory with 11 corollaries and novel implications for assessment and psychotherapy. Our focus here, however, is on the person-as-scientist metaphor itself and its unexplored implications for psychological theory more generally. In addition to its emphasis on the primary sense-making orientation of human beings, personal construct theory considered as a whole reveals several other characteristics of scientists – the acting person of science practice – to be implied by Kelly's scientist metaphor. All of these are important to consider in our framing of science as psychology.

1. The scientist is an actor, not a reactor. Kelly's intended point of contrast is one of the person as a "propelled" organism (Kelly, 1963, p. 5), a conception of person emerging in the context of a mid-20th-century psychology dominated by an emphasis on behavioral shaping through stimulus–response contingencies. Kelly's emphasis is on the intentional, goal-directed activity of persons. Persons "not only construe their alternatives, but they construe their criteria for choosing between them (Kelly, 1969, p. 84). However, Kelly's emphasis on the active, intentional orientation of humans (and he does not exclude animals from this orientation) is not intended to imply unbounded freedom or limitless opportunity for construction. Clear constraints are imposed by the physical and social world. Moreover, they are imposed by our own interpretive system itself (existing construct system), which implicates memory. That is, the sense we have made of events through our experience to date imposes direction and limits: "All thinking is based, in part, on prior convictions" (1963, p. 6):

Man's thinking is not completely fluid; it is channelized. If he wants to think about something he must follow the network of channels he has laid down for himself, and only by recombining old channels can he create new ones. These channels structure his thinking and limit his access to the ideas of others. We see these channels existing in the form of constructs (1963, p. 61).

Yet the "choice corollary" of PCT ensures that the person/scientist "places relative values upon the ends of his dichotomies. Some of the values are quite transient and represent merely the convenience of the moment. Others are quite stable and represent "guiding principles" (1963, p. 62).

As noted, the centrality of action in Kelly's thought invites comparisons of personal construct theory to Dewey's account of thinking (Dewey, 1910, 1938b), and indeed, Kelly describes the psychology of personal constructs as following Dewey in its emphasis on the anticipatory nature of behavior and the active formation and testing of ideas in thinking. Construing is *activity*, with predictive efficiency the marker of the extent to which a situation is transformed from problematic to temporarily resolved. Accompanying the emphasis on action, Kelly offers a process or movement account of personhood, in keeping with the depiction of the universe as a series of temporally structured events: "For our purposes the person is not an object which is temporarily in a moving state but is himself a form of motion" (Kelly, 1963, p. 48); elsewhere, he wrote,

> Life itself could be defined as a form of process or movement. Thus in designating man as our object of psychological inquiry, we would be taking it for granted that movement was an essential property of his being, not something that had to be accounted for separately. We would be talking about a form of movement – man – not something that had to be motivated (1969, p. 80).

The activity of construing requires revision and readjustment, for the scientist Kelly envisions has integrity and follows sound methodology:

> A scientist formulates a theory – a body of constructs with a focus and a range of convenience. If he is a good scientist, he immediately starts putting it to test. It is almost certain that, as soon as he starts testing, he will also have to start changing it in the light of the outcomes. Any theory, then, tends to be transient. . . . At best it is an ad interim theory (Kelly, 1963, p. 14).

The emphasis on revision of theories is at the core of what Kelly identifies as personal construct theory's philosophical position: "Since an absolute construction of the universe is not feasible, we shall have to be content with a series of successive approximations to it. . . . We assume that all of our present interpretations of the universe are subject to revision or replacement" (1963, p. 15). Constructs are "patterns that are tentatively tried on for size"

(p. 9). This principle of *constructive alternativism*, which is a core of Kelly's philosophy as a whole and particularly of his understanding of scientist, is also the basis of therapeutic process. Alternative constructions of one's self and life events are always ready to hand, and transformations follow from new constructions: "No one needs to be a victim of his biography" (p. 15).

In our studies of research scientists in biomedical engineering laboratories, we have characterized the laboratories themselves as *agentive learning environments*, thereby emphasizing that features of the socio-cognitive and material environment contribute to the production of learners able to act strategically to plan and coordinate their learning tasks and problem solving. This emphasis on the agentive *environment* underscores not only Kelly's emphasis on the active rather than reactive scientist but also the contribution of the physical and social configuration to maximizing the scientist's potential for active engagement.

2. The scientist is engaged *dialectically*; scientific reasoning is inherently *dialectic*. Kelly's student, Joseph Rychlak, has made emphasis of this point central to his characterization of personal construct theory (e.g., Rychlak, 1990). Constructive activity and the problem solving that accompanies the revision of constructs are characterized by dynamic opposition. At a most basic level, every construct implies a contrast: "In its minimum context a construct is a way in which at least two elements are similar and contrast with a third" (Kelly, 1963, p. 61). Dichotomy is "an essential feature of thinking itself" (p. 62). Kelly characterizes the dichotomy corollary of personal construct theory as follows:

> Having chosen an aspect with respect to which two events are replications of each other, we find that, by the same token, another event is definitely not a replication of the first two. The person's choice of an aspect determines both what shall be considered similar and what shall be considered contrasting (p. 59).... What we mean is that there is an aspect to A, B, and C which we shall call z. With respect to this aspect, A and B are similar and C stands in contrast to them (p. 60).... In its minimum context a construct is a way in which at least two elements are similar and contrast with a third (p. 61).

Thus constructs cannot be evaluated without analysis of the implicit contrast class, that from which the events seen to be similar are distinguished. This process is easiest to appreciate in relation to linguistic phenomena. Consider this commonplace example with political overtones: The meaning and value of "elite" differ radically depending on whether the intended contrast is "second-rate" or "grounded in common sense." To offer an example from our data, an engineering researcher's construal of cells as "happy" implies a contrast with another state. Understanding the contrast intended is an important task in analyzing the meaning of discursive strategy employed so widely in the laboratories we investigated.

3. The scientist construes at *different levels of awareness*. Because of Kelly's distaste for the notion that a person is propelled toward determined courses of affect and action, he resisted reference to psychodynamic conceptions of an agentive unconscious mind. Nevertheless he was convinced that much of our sense-making takes place below the surface of awareness and that we can neither articulate nor even appreciate much of the structuring of our experience accomplished in our constructive activity. Similarly, "construing is not to be confounded with verbal formulation" (1963, p. 51). Kelly's assumption is that construing takes place on a continuum of awareness ranging from very low to very high awareness. Low levels of awareness are indicated by a sudden reaction or even autonomic system activation or a digestive disturbance without accompanying insight into the triggers of the reaction.

Kelly's depiction of constructing as discriminating activity made room for preverbal constructions that are the most formative and resistant to revision, those constructs that he considered "core." A primary distinction between safe and unsafe events is an example of this level of sense-making; core constructs are formatively related to the infant's survival. As life progresses these core constructs may be layered with increasingly sophisticated words and images, yet continue to orient in originative ways: "We recognize that the psychological notion of construing has a wide range of convenience, which is by no means limited to those experiences which people can talk about or those which they can think about privately" (1963, p. 51).

4. The scientist is *reflexive*. Although construing is not purely "conscious" activity, we clearly are able to attend to the contents of our own experiencing, to reflect on our own construing. This ability to reflect is pivotally related to the transformative power of PCT, enabling the scientist (and client) to consider construct systems deliberately and actively subject them to test or alteration. Constructs are not uniformly available to attention, of course. Yet the potential to bring one's constructs into focus is key to conceptual change in science and psychotherapy. Fransella (1995) characterizes "the idea of reflexivity" as "central to the whole of Kelly's thinking" (p. 60) and psychotherapy as a means by which a client will be "learning to become a more competent personal scientist" in the process (p. 61). Our own research group has identified *agentive learning* as essential to a researcher's progress and ultimate success in the biomedical engineering labs under investigation.

Agentive learning as we have described it is reflexive and deliberate learning; it entails critical reflection on one's present tasks in relation to one's overarching problem focus and continual purview of specific strategies in relation to the overall learning goals.

5. The scientist acts *holistically; construing is holistic activity*. The imposition of order and sense is an act of *persons*, not a function of processes or mechanisms or any part of the person in isolation. For this reason Kelly stringently avoids reification of the traditional categories of psychological study, including motivation, and the implicit artificial dichotomies imposed by traditions of psychological discourse: mind and body, intellect and emotion, cognition and experience. He is clearest on this point when responding to the generalized critic of personal construct psychology who interprets the theory as too intellectualized:

> In an effort to understand what we are talking about a listener often asks if the personal construct is an intellectual affair. We find that, willy-nilly, we invite this kind of question because of our use of such terms as thought and thinking. Moreover, we are speaking in the terms of a language system whose words stand for traditional divisions of mental life, such as "intellectual."
>
> Let us answer this way. A construct owes no special allegiance to the intellect, as against the will or the emotions. In fact, we do not find it either necessary or desirable to make that classic trichotomous division of mental life. After all, there is as much that is "emotional" in those behaviors commonly called "intellectual" and there is so much "intellectualized" contamination in typical "emotional upheavals" that the distinction merely becomes a burdensome nuisance. For some time now we have been quite happy to chuck all of these notions of intellect, will, and emotion, and so far, we cannot say we have experienced any serious loss (1969, pp. 87–88).

Fransella thus describes Kelly's theory as free from "segments" (1995, p. 115).

Of course, psychology's history has produced a strong showing of "whole person" accounts as alternatives to the field's compartmentalized and fragmented norm. Yet despite the good intentions of Third Force critiques, some humanistic enthusiasts have taken on the appeal to holistic accounts (and the whole person) as a mantra and end in itself, as a rationale for avoiding additional theorizing. Kelly, however, is firmly rooted in a rigorous tradition of personalism by which persons are understood as metaphysical primaries (Brightman, 1943), as well as in pragmatist responses to the "new physics" that place this agentive process at the center of science (Dewey, 1929; Watson, 1938). Thus the whole person framework by which Kelly understands the scientist requires new descriptive tools and new units of analysis, but is in no way intended as a celebration of irrationality. The holistic framework is embraced to emphasize that the active choice entailed in the construction of alternatives is not a function of an isolated faculty: "When one makes a choice he involves *himself* in the selection. Even if the choice is no more than a temporary hypothesis explored in the course of solving a mathematical problem or in looking for a lost screwdriver, he must perceive *himself* as being modified through the chain of ensuing events" (Kelly, 1969, p. 65).

Similarly, the inseparability of personal meaning-making from the social context in which the meaning-making occurs is an important aspect of the theory, despite historical criticisms that PCT is insufficiently social and particularly lacks appreciation of the social and cultural forces by which personal constructs are produced (e.g., Proctor & Parry, 1978). A close reading of *The Psychology of Personal Constructs* reveals personal construct formation and revision to occur always within a given social realm and to be constrained by it. *Sociality* is thus among personal construct theory's central corollaries. The sociality corollary includes Kelly's assumption that "the subsuming of other people's construing efforts" is "the basis for interpersonal interaction" (Kelly, 1969, p. 95).[4] Persons vary in the extent to which they are able to subsume another's construct system and in the extent to which their construct system is subsumable. Yet our very survival is tied to this ability to grasp another's constructs, and this ability is evident in our most basic normative activities:

> The orderly, extremely complex, and precise weaving of traffic is really an amazing example of people predicting each other's behavior through subsuming each other's perception of a situation. Yet actually each of us knows very little about the higher motives and complex aspirations of the oncoming drivers, upon whose behavior our own lives depend (1969, p. 97).

Thus our own constructive activities are at every level bound to the web of norms and expectations that establish the social order in its manifold variations. Kelly understands culture in terms of "similarity of expectations" (p. 93) – both what one expects others to do and what one believes one is expected to do – leading persons in the same cultural group to construe their experience along similar "channels" or directions. Hence there is for Kelly no distinct "social process" to *interact* with our basic sense-making, construing practices.

---

[4] This invites an obvious link to the literature on "theory of mind" (e.g., Wellman, 1990).

Here Kelly's intertwining of personal construction with normative order and cultural specificity complements Nersessian's call for a view of the cognitive and cultural realms as facets of the same complex system, rather than as distinct systems that interact (or in which the social functions as confounding influence). The cognitive-cultural system, as she has called it (Nersessian, 2005), is the background assumption against which model construction, manipulation, and dissemination are understood in the engineering laboratories under her investigation.

6. The scientist is an inveterate *innovator*. In contrast to the plodding, mechanical picture of science emerging from some critics of scientism, the most central feature of PCT, and one that is key to Kelly's understanding of the person as scientist, is the inherent potential for conceptual innovation, for revision of even long-standing constructions of reality in the face of new experience and interpretative schemes. Such potential, summarized by the principle of "constructive alternativism," is similarly the basis for change in therapy and for human adaptation and fulfillment more broadly: "What I am saying is that it is not so much what a man is that counts as what he ventures to make of himself" (Kelly, 1964, p. 157). The scientist-researcher as innovator is a model for the inherent potential of human beings to generate alternatives for living, even in the most adverse circumstances.

Implications of the Person-as-Scientist Metaphor

Taken together, features of "the scientist" emerging from the corollaries of personal construct theory have important implications for understanding both science and persons. For George Kelly, the experiencing, construing person, acting indivisibly within social and material contexts, is the starting point and center of science process. There is at the same time something fresh and something well worn and familiar about this emphasis. It is freshest when considered against the contemporary backdrop of science studies, which, although extraordinarily fertile, have unwittingly bequeathed a set of misleading divisions, not unlike the divisions evident within psychological science more specifically. As noted in relation to the integration problem we identified in our introductory chapter, the effect is a virtual impasse in science studies around the question of how science is to be understood, whether as chiefly a social *or* cognitive phenomenon (Longino, 2002; Nersessian, 2005).

Moreover, there is a dimension of science practice that is not adequately captured by social *or* cognitive dimensions: a "something else" related to the scientist's particularity of personhood.

Science and Personhood

The primacy of the *personhood* of the scientist, in contrast with *personality* (as a collection of traits), is emphasized in a few intriguing contributions to the psychology of science. It is in relation to these works that the familiar aspect of George Kelly's depiction of the scientist becomes apparent. These texts are infrequently mentioned in science studies at large, in part because their "fit" within the conventional divisions of science studies and even the psychology of science is not immediately evident. Polanyi's *Personal Knowledge* and Maslow's *The Psychology of Science* were mentioned earlier in this chapter, but there are other important examples with philosophical as well as psychological merit. A vibrant but little-known work by D. L. Watson was published in 1938 with the title *Scientists are Human* as one volume in a collection entitled *The Library of Science and Culture*. With a foreword by John Dewey, Watson's work affirms "the scientist" to be the chief problem of science, claiming that "central to any estimate of the nature of scientific truth and its value for humanity are (1) an understanding of the psychological constitutions of the investigator and (2) an understanding of the social forces which produce him, encourage or oppose him, and transmit or ignore his work," and these two understandings are viewed as complementary.

Like Kelly, Watson grounds his affirmation of the importance of the scientist *to science* in implications he draws from the new physics of his time, calling the means by which the scientist and his or her science are seen as – in a new sense – *inseparable* to be the most significant fruit of relativity theory. The book *Scientists Are Human* is thus devoted to unpacking what Watson calls the "intricate personal and social matrix out of which discoveries arise" by detailing examples of scientists' achievement of coherence in what Bridgman terms the "penumbra of uncertainty" (quoted in Watson, p. 16). In the book's first half, Watson acknowledges inborn dispositional or stylistic contributions to the production of successful scientists, such as a talent for original thought interacting with a taste for solitude, giving historical examples ranging from Newton to Faraday. However, he tempers the appeal to dispositional or personality factors with an appeal to social factors, even to the extent of producing the conditions wherein scientific interest and temperament are made possible and rewarded:

> Let us imagine that certain desirable temperamental tendencies have been built into the character of the embryonic scientist by the social and physiological circumstances of his birth, growth, and upbringing.

Now comes a trial of strength between these tendencies and the organized customs of the adult society. The searching activities of the scientific man provide, in the first place, an undifferentiated mass of facts and fancies. From this a socially acceptable interpretation is gradually sifted out. The social forces select certain of the scientist's abilities and aptitudes for approval. The social organization acts as a filter on the scientific results which are candidates for recognition. Where the belief and practice of the society are sound, this process leads to scientific truth (Watson, 1938, p. 50).

The "beliefs, customs, prejudices, and institutions of society all influence the science which it fosters" (p. 49) and are an essential part of what it means to understand the scientist as human.

The second half of *Scientists are Human* features a 90-page chapter entitled "On the Similarity of Forms and Ideas" in which Watson looks to Gestalt psychology, as Polanyi was soon to do, to defend his assertion that the "unverbalized discerning of intrinsic resemblances" or similarities of form or idea are "the chief kind of knowledge we have" and thus the basis of science and scientific knowledge. Although the Gestaltists emphasized the active, organizing powers of perception, Watson recognizes anticipations of Gestalt principles in Aristotle's notion of active intellectual organization, Locke's rational powers, and other historical examples. For Watson, what the mind adds to nature "to organize what it sees" includes contributions of the psyche of which the scientist is not aware, unplanned operations such that new configurations seem to jump out in problem solving. Thus although Watson's aim was in part to expose the fallibility of science introduced by the vulnerability of the human knower, he calls the unconscious "a rich storehouse of potential insights into the impersonal behavior of the *non-human world*, and therefore an aid for the problems of science" (1938, p. 189, emphasis original). His implication that this subconscious is structured both by the social world and by the basic operations of apprehending similarities of form and idea links what might otherwise look like disconnected portions of his text. Dewey's foreword calls this a "well-grounded plea for a recognition of the inherent likeness of aesthetic response and of artistic creativity with genuine scientific procedure" (p. x).

Watson's characterization of science bears a marked similarity to Michael Polanyi's depiction of scientific discovery as "an extremely delicate and personal art which can be little aided by any formal percept" (1946/1964, pp. 33–34). Particularly important in Watson's insights is the implication that the social world is inseparably a part of the person at the center of science.

More recently, Michael Mahoney and Ian Mitroff, with quite different styles and methods, have emphasized the paramount importance of the personhood of the scientist for understanding science. Obviously drawing a contrast to the objectivity so long associated with the adoption of a common method, Mahoney and Mitroff focus on the contributions of *subjectivity* to the practices and products of science.

Mitroff was mentioned earlier in connection with his derision of the "storybook" view of science prominent in science education sources under his review. His critique of this view is posed as a "social psychology of research" (p. 20), a "book about how science actually gets done" (p. 2) with broad application to philosophy of science. Unlike many sociological analyses of science, however, the critique foregrounds the passion, emotional commitment, and personal investment of scientists as contributing influences in the insights achieved and decisions made. It is based on an analysis of interviews with 42 scientists associated with the Apollo lunar missions of the late 1960s, including NASA scientists, university scientists, and other government or private industry scientists. Though open-ended in format, each interview asked for reflection on the nature of reaching the moon as a physical problem and the nature of science as an institution and as a means of producing knowledge. Perhaps even more interesting than his own rejection of the storybook account of science is the response of the scientists Mitroff interviewed to the depiction of scientists as objectively and disinterestedly pursuing truth:

> All of the 42 scientists interviewed indicated in one way or another that they thought the notion of the objective, disinterested scientist was naïve. The vocal and facial expressions that accompanied the verbal responses were extremely revealing and important. They ranged all the way from mild humor and guffaws to extreme annoyance and clear expressions of anger. The respondents felt that the only people who took the idea of the objective, disinterested scientist literally and seriously were the general public or beginning science students (Mitroff, 1974, p. 64).

Most interesting, however, is the third category of question Mitroff included: "issues that are related to scientists as persons." He explains, "Here the question is how the scientists felt about one another in personal terms as peers and as colleagues" (p. 42). Questions include, for example, "What in your opinion distinguishes those scientists who formulate theories from those who do not? Are there any scientists whom you regard as highly committed to or even biased in favor of, their pet theories?" (p. 43).

Mitroff's analysis of the contribution of personal factors to science focuses on variations in style, dispositional and cognitive, that contribute to what are essentially different ways of doing science: "Radically different observers or types of scientists bring radically different types of presuppositions with them to the field of scientific observation" (p. 257). Despite the reference to types, Mitroff's analysis does not reflect standard trait theory, but rather a distinction between rational styles not unlike that identified by William James as "forms of mental temper" (1896, p. 90). Relating his findings to the insights of Kuhn (1962), Feyerabend (1975), and Toulmin (1972), Mitroff explains,

> It seems to me that the data of this study suggest that in every social system there are those kinds of individuals (types) who have a compulsive need to make revolutions, to disagree, as strongly as possible, with established ways of thinking – paradigms, if you prefer. These individuals have an almost consuming need to produce radical ideas and theories counter to those currently in existence if not in vogue. They seem to need to go out of their way to produce extremely novel ways of looking at old phenomena. However the data also suggest that there are also those kinds of individuals who have a compulsive (security) need to preserve continuity with old established ways of thinking, to differ as little as possible from the tried and true. Complete disagreement or agreement, rather than being actual states of affairs, are instead states of mind, attitudinal Ideals, or, in Toulmin's terms, the divergent "disciplinary aims" of radically distinct types of men (p. 261).

Mitroff draws out the implications of this finding of different types of observers to critique the notion of emotional neutrality and impersonality as characteristic of the nature of science. Mitroff's conception of the psychological is clearly one in which social, cognitive, and dispositional factors are irreducibly configured into the nature and practice of rationality.

Mitroff's description of the storybook view of science is paraphrased in the first chapter of Michael Mahoney's *Scientist as Subject: The Psychological Imperative*, the first edition of which was published in 1976. Like Mitroff, Mahoney derides the notion of emotional neutrality as not only infeasible but also undesirable. Mahoney's text is a collection of essays on topics such as rites of passage in graduate training, the politics of university life, and the demands of publication. The inevitable emotional response of the scientist confronted with these historically sanctioned institutional structures infuses problem solving and discovery with an amalgamation of social and emotional energy. Mahoney thus wages a full-scale attack on the assumption that science is driven principally by disinterested rationality and argues that an understanding of its emotional and social foundations is necessary for a full understanding of the products and truth claims of science.

Mahoney's insights into the passions and blind ambitions driving the day-to-day operations of science are offered with the intent of informing more fully and richly the nature of the scientist, not as a personality type but as an exemplar of passions and foibles characteristic of human beings. Greater understanding of scientists' humanity, here equated primarily with emotionality as it infuses and directs reason, is intended to give us a more accurate understanding of science:

> Our relative ignorance about homo scientus is not, in my opinion, a harmless omission which simply disturbs the esthetic balance of our current knowledge. The oversight is not benign. Our continued neglect of the scientist could well be the most costly blunder in the history of empiricism. We can hardly hope to have much confidence in the products of science if we choose to remain ignorant of the limitations imposed by its human embodiment (Mahoney, 2004, p. xxx).

Two features of the accounts of science offered by Watson, Mitroff, and Mahoney (as well as Polanyi, Maslow, and Kelly) are remarkable and too infrequently emphasized in contemporary science studies. The first is a subject-centered account of science, with the understanding that subject centered is not synonymous with "subjective" in the sense of irrational, arbitrary, solipsistic, or *merely* personal. The implication is, instead, that making observations, drawing inferences, testing hypotheses, developing theories, and modeling reality are *acts of persons*, and therefore a full under-standing of the nature of persons with all of the features of their embod-iment and sociality is necessary to an accurate account of their rational powers.

This seems so obvious that it is difficult to imagine how an alternative could be taken seriously. Yet the *language* surrounding both radically social and traditionally cognitive accounts is frequently suggestive of such an alternative.

Here we single out a recent example which describes the development of the scientific attitude historically:

> Evolution did not produce brains so they could do systematic and explicit math and science, but it did evolve a sophisticated central nervous system that organizes and interprets sensory information and is able to reflect upon experiences and put thought between impulse and behavior. In the

process of organizing and interpreting sensory input, the brain recognizes patterns, makes causal connections, and forms expectations and predictions (hypotheses; Feist, 2006a, p. 217).

On the face of it there is nothing wrong with this account. However, it contains an implication that *brains* in isolation are doing the interpreting, expecting, and predicting in such a way that leads to knowledge, even though we assume that was not the intention. That is, brains are assumed to be the kind of entity capable of performing the activities of organizing, expecting, and predicting.

The problem here is best appreciated in the context of Bennett and Hacker's description of the "mereological fallacy," the "mistake of ascribing to the constituent parts of an animal attributes that logically apply only to the whole animal" (2003, p. 73). In this category Bennett and Hacker include the practice of attributing to the brain psychological attributes and abilities, by which the brain "knows things, reasons inductively, and constructs hypotheses" (p. 69), a practice widely adopted in cognitive neuroscience "without argument or reflection" (p. 72). Yet, as they note, "Human beings, but not their brains, can be said to be thoughtful or thoughtless . . . can be said to make decisions or to be decisive" (p. 72). Whenever we talk in terms of brains doing science instead of persons, there is then a *category mistake* (Ryle, 1949), proliferated in many descriptions of the cognitive bases of science. The problem is in no way limited to cognitive accounts. It applies equally to radically social accounts of science that focus on the systemic and macro-level processes responsible for knowledge production to the exclusion of the experiencing scientist (e.g., Latour, 1987).

Yet what should be included in an account of subjectivity and how best to analyze its contribution? This is a question first of content and second of method. In terms of content or focus, Watson, Mitroff, and Mahoney in various ways highlight the impact of temperament, commitment, and institutional forces on the intricacies of problem solving in science. We thus have included in this volume a chapter on emotion and motivation and analyze the subtle ways in which evidence of emotion is embedded even in descriptions of procedure and equipment. Yet subjectivity cannot be adequately understood without other levels of description or other foci. Certainly the visualization and analogy practices so closely linked to perceptual and imaginative powers must be considered part of the scientist's subjectivity. Yet so must efforts to achieve coherence in reasoning that include positioning one's cognitive achievement in relation to others, both locally and in the field at large. So, too, must the scientist's physical constitution and the historical categories by which this embodiment is understood and reacted to by others: Thus an analysis of gender and race "performances" in the practice of science is an important component of our analysis. Finally, and relatedly, we must consider subtle shifts in identity formations triggered by encounters with other persons with whom one practices science or even with the objects and artifacts of one's practice. We have thus focused on sense-making and social identity as central to a person-centered account of science practice.

More accurately, perhaps, what is required is not a "subjective account of science" so much as an account that calls into question the viability of a sharp subject/object distinction in the first place. This is the strategy adopted by the hermeneutic tradition as it has been incorporated into the philosophy of science. Although this tradition is far too comprehensive to adequately review within these pages, the relation of this tradition to a practice framework (in contrast to abstracted processes) and to the activity of the scientist is articulated clearly in McGuire and Tuchanska's comprehensive *Science Unfettered*:

> Being a subject of scientific activity is to participate in a particular subpractice that belongs to the entire practice of certain historical societies. This participation is a form of being-together-in-the-world: through participating in research an individual takes part in social relations that form the structure of being-together and with other scientists and with the life-world of science (2000, p. 150).

These authors note, however, that "[t]hey are highly individualized and personalized exchanges (mainly discursive), in which all of the varying idiosyncrasies come into play among participants." They further note that the need for creativity in scientific practice necessitates that "scientific communities embrace scientists as persons," resulting in what they term "participation within an ontic circle of individuality and conformity" (p. 150).

We would only note that this view of science as participation, and the view of participation as necessarily implicating both the creative contributions of particular embodied persons and the many levels of social organization within which they are expressed, is not confined to the "continental tradition" per se. As we have attempted to demonstrate through this brief review, the subject/object (scientist/world) dichotomy is called into question and challenged in various ways by accounts of scientific practice as "thoroughly American" as those of John Dewey, George Kelly, Abraham Maslow, Ian Mitroff, and Michael Mahoney, as well as by many contemporary efforts in cognitive science. If it is the case that these perspectives reflect implicit saturation with continental commitments, so be it. However, we find it

unhelpful to inscribe rigid lines of demarcation here, as they contribute
to the too frequent tendency for scholars to line up on different sides. We
hope to have brought some needed attention to the conceptual advantages
of regarding the acting person as a unit of analysis in the psychology of science precisely because it requires integrated accounts of particular scientific
practices.

As we have also attempted to make clear, our emphasis on the acting
person is not meant to convey the impression that science is merely a
"subjective" enterprise. For Polanyi, D. L. Watson, Mitroff, and Mahoney,
appeal to the influence of personal influences in science does not imply
that science is an irrational or arbitrary enterprise, nor does it warrant
avowedly relativistic conclusions. Neither is the framework "individualistic"
in the traditional sense, for at every level the activity of the acting person is
situated in structuring, normative social contexts and inherited conceptual
and social categories. Our interest is in offering a more rounded account
of science process, to work toward an account of rationality in science
that includes the full measure of embodiment, intentionality, particularity
(style), sociality, and cognitive activity mandated by the very concept of
person.

SCIENCE AS PSYCHOLOGY

We repeatedly have stated that our interest in the acting person extends
beyond the psychology of science into the broader realm of general psychology, which is also in need of integration. This assertion, too, finds historical
precedent, as in Dewey's comparison of the scientist and the "man in the
street":

> In the natural sciences there is a union of experience and nature which is
> not greeted as a monstrosity. . . . The investigator assumes as a matter of
> course that experience, controlled in specifiable ways, is the avenue that
> leads to the facts and laws of nature. He uses reason and calculation freely;
> he could not get along without them. But he sees to it that ventures of this
> theoretical sort start from and terminate in directly experienced subject-
> matter. . . . And this experienced matter is the same for the scientific man
> and the man in the street (Dewey, 1929, p. 2a).

In our discussion of biomedical engineering researchers we have sought
to demonstrate the connection between various aspects of science practice
and general psychology by organizing chapters around some traditional

foci of general psychology: cognition (problem solving), emotion, identity (positioning), sociality (race and gender enactment), and learning. Yet here we offer summative comment. What does an integrated account of the sense-making and social identity of practicing scientists offer toward a general psychology? On what grounds might we draw extensions that are informative beyond the laboratories we have considered in this text, particularly when many psychologists assume that qualitative analysis does not lend itself to generalizations beyond the local group of participants analyzed? The claim that we can offer our analysis as a contribution to general psychology takes inspiration from George Kelly's person-as-scientist metaphor. For Kelly the acting scientist provides the most accurate and vivid representation of the sense-making and identity-forming activities of persons seeking to understand and manage their lives, for which reason it serves as the foundation for his approach to psychotherapy.

Yet the implications of George Kelly's person-as-scientist metaphor extend beyond the project of trying to understand science practice, just as they transcend clinical practice contexts. To understand these implications requires some attention to the nature and purpose of metaphor more broadly. Mary Hesse's (1966) classic work on models and analogies in science includes a final chapter on the explanatory function of metaphor. Offering an expository analysis of Black's interactionist view of metaphor, Hesse emphasizes,

> The metaphor works by transferring the associated ideas and implications of the secondary to the primary system. These select, emphasize, or suppress features of the primary; New slants on the primary are illuminated; The primary is seen through the frame of the secondary.... [But also] *The same applies to the secondary system*, for its associations come to be affected by assimilations to the primary; the two systems are seen as more like each other... men are seen to be more like wolves after The Wolf metaphor is used, and wolves seem to be more human. Nature becomes more like a machine in the mechanical philosophy, and actual, concrete machines are themselves seen as if stripped down to their essential qualities of mass in motion (Hesse, 1966, pp. 162–163, emphasis added).

For the person-as-scientist metaphor, both "systems" (person and scientist) are mutually informative. Thus by understanding more about persons we understand more about scientists (hence, science), and understanding more about scientists yields a similar transfer to the understanding of persons.

As a caveat, we cannot claim to have studied the person of science as fully and deeply as is ultimately required. We have tried to address a major subset of traditional psychological foci (categories), but have not been exhaustive in this regard. In the confines of one book it is not possible to devote sufficient attention to all the important dimensions of personhood as they are implicated in and by science practice. One glaring omission is the role and function of value commitments in science. As we have noted, there is an enormous literature on science and values that we have left virtually untouched. Values are hinted at as we take up other topics. We have noted, for instance, some points at which questions about values overlap with questions about the function of emotional commitments in science and with identity. Yet we decided we could not adequately address the big questions around science and values within the scope of this text. What we can say is that values permeate our interviews as they permeate the research of these biomedical engineers. If we look at values in science with the useful categories Helen Longino (1990) has developed – "constitutive" values and "contextual" values – as with all science, the engineering sciences possess values pertaining to what are acceptable practices; unquestionably, at the very least, research agendas in the biomedical sciences and engineering are embedded in values emanating from the wider social and cultural context.

The fundamental point of the collective effort presented here is that to understand science requires a study of scientists engaged in actual practices, which includes both observing what they do and listening to what they say they do. At every step this requires sense-making and identity formations for the student of the scientist no less than for the scientists themselves.

REFERENCES

Adams, F., & Aizawa, K. (2008). *The bounds of cognition.* Oxford: Blackwell.

Adler, A. (1992). *Understanding human nature* (C. Brett, Trans.). Oxford: Oneworld Publications. (Original work published 1927)

Alac, M., & Hutchins, E. (2004). I see what you are saying: Action as cognition in fMRI brain mapping practice. *Journal of Cognition and Culture, 4*(3–4), 629–662.

Alasuutari, P. (1995). *Researching culture: Qualitative method and cultural studies.* Thousand Oaks, CA: Sage.

Alcoff, L. (2006). *Visible identities: Race, gender and the self.* Oxford: Oxford University Press.

Alloy, L. B., & Abramson, L. Y. (1988). Depressive realism: Four theoretical perspectives. In L. B. Alloy (Ed.), *Cognitive processes in depression* (pp. 223–265). New York: Guilford.

Alper, J. (1993). The pipeline is leaking all the way. *Science, 260,* 409–411.

Alsop, R., Fitzsimmons, A., & Lennon, K. (2002). *Theorizing gender.* Cambridge: Polity Press.

Anscombe, G. E. M. (1966). *Intentions.* Oxford: Blackwell.

Anderson, M. (2003). Embodied cognition: A field guide. *Artificial Intelligence, 149,* 91–130.

Arnold, M. B. (1960). *Emotion and personality: Vol. 1. Psychological aspects.* New York: Columbia University Press.

Atkins, K. (2005). *Self and subjectivity.* Oxford: Blackwell.

Auxier, R. (2002). Preface. In R. Auxier & L. E. Hahn (Eds.). *The philosophy of Marjorie Grene* (pp. xvii–xx). Library of Living Philosophers Series. Peru, IL: Open Court Publishing.

Baars, B. J. (1986). *The cognitive revolution in psychology.* New York: Guilford.

Bannister, D., & Mair, J. M. M. (1968). *The evaluation of personal constructs.* London: Academic Press.

Baron-Cohen, S., Bolton, P., Wheelwright, S., Short, L., Mead, G., Smith, A., & Scahill, V. (1998). Autism occurs more often in families of physicists, engineers, and mathematicians. *Autism, 2,* 296–301.

Barsalou, L. (2008). Grounded cognition. *Annual Review of Psychology, 59,* 617–645.

Bechara, A. (2004). The role of emotion in decision-making: Evidence from neurological patients with orbitofrontal damage. *Brain and Cognition*, *55*(1), 30–40.

Beck, A. (1970). *Depression: Causes and treatment*. Philadelphia: University of Pennsylvania Press.

Becvar, A., Hollan, J., & Hutchins, E. (2008). Representing gestures as cognitive artifacts. In M. S. Ackerman, C. Halverson, T. Erickson, & W. A. Kellogg (Eds.), *Resources, co-evolution, and artifacts: Theory in CSCW* (pp. 117–143). New York: Springer.

Bedford, E. (1962). Emotions. In V. C. Chappell (Ed.), *The philosophy of mind* (pp. 110-126). Englewood Cliffs, NJ: Prentice-Hall.

Bennett, M. R., & Hacker, P. M. S. (2003). *Philosophical foundations of neuroscience*. Oxford: Blackwell.

Ben-Ze'ev, A. (2000). *The subtlety of emotion*. Cambridge, MA: MIT Press.

Berkson, W. & Wettersten, J. (1984). *Learning from Error: Karl Popper's Psychology of Learning*. LaSalle, IL: Open Court.

Bickhard, M. H. (2008). *Are You Social? The Ontological and Developmental Emergence of the Person*. In U. Mueller, J. I. M. Carpendale, N. Budwig, & B. Sokol (Eds.) *Social Life and Social Knowledge* (pp. 17–42). New York: Taylor & Francis.

Billig, M. (2002). Henri Tajfel's cognitive aspects of prejudice. *British Journal of Social Psychology*, *41*, 171–188.

Bloor, D. (1976). *Knowledge and social imagery*. London, UK: Routledge & Kegan Paul.

Bourdieu, P. (1990). *The logic of practice*. Stanford: Stanford University Press. (Original work published in French as *Le sens practique*, 1980)

Bowlby, J. (1991). *Charles Darwin, a new life*. New York: Norton.

Brentano, F. (1995). *Psychology from an empirical standpoint* (P. Simons, Trans.). New York: Routledge. (Original work published as *Psychologie vom empirischen Standpunkt*, 1874)

Bridgman, P. (1950). *Reflections of a physicist*. New York: The Philosophical Library.

Brightman, E. S. (1943). Personality as a metaphysical principle. In E. S. Brightman (Ed.), *Personalism in theology* (pp. 40–63). Boston: Boston University Press.

Brockmeier, J. (2009). Reaching for meaning: Human agency and the narrative imagination. *Theory & Psychology*, *19*(2), 213–233.

Brown, R. (2000). Social Identity Theory: past achievements, current problems and future challenges. *European Journal of Social Psychology*, *30*, 745–778.

Brown, B. (2004). Discursive identity: Assimilation into the culture of science and its implications for minority students. *Journal of Research in Science Teaching*, *41*(8), 810–834.

Brown, J. S., Collins, A., & Duguid, P. (1989). Situated cognition and the culture of learning. *Educational Researcher*, *18*, 32–42.

Brown, R. (2002). Henri Tajfel's 'Cognitive aspects of prejudice' and the psychology of bigotry. *British Journal of Social Psychology*, *41*, 195–197.

Brown, S. (2007). Intergroup processes: Social identity theory: Introduction. In D. Langdridge & S. Taylor (Eds.), *Critical readings in social psychology* (pp. 133–135). Maidenhead, UK: Open University Press.

Burke, P. J., & Stets, J. E. (2009). *Identity theory*. New York: Oxford University Press.

Burns, L., Einaudi, P., & Green, P. (2009). *S&E graduate enrollments accelerate in 2007; Enrollments of foreign students reach new high* (NSF Report 09-314). Retrieved from National Science Foundation, Division of Science Resources Statistics, http://www.nsf.gov/statistics/infbrief/nsf09314/nsf09314.pdf.

Burrelli, J. (2008). *Thirty-three years of women in S&E faculty positions* (NSF Report 08–308). Retrieved from National Science Foundation, Division of Science Resources Statistics, http://www.nsf.gov/statistics/infbrief/nsf08308/.

Butler, J. (1990). *Gender trouble and the subversion of identity*. New York: Routledge.

Butler, J. (1993). *Bodies that matter: On the discursive limits of "sex."* New York: Routledge.

Carlone, H., & Johnson, A. (2007). Understanding the science experiences of successful women of color: Science identity as an analytic lens. *Journal of Research in Science Teaching, 44*(8), 1187–1218.

Carey, S. (1985). *Conceptual Change in Childhood*. Cambridge, MA: MIT Press.

Carnap, R. (1935). *Philosophy and logical syntax*. London: K. Paul, Trench, Trübner & Co., Ltd.

Carruthers, P., Stich, S., & Segal, M. (2002). *The cognitive basis of science*. Cambridge, UK: Cambridge University Press.

Chandrasekharan, S., & Nersessian, N. J. (2008). Counterfactuals in science and engineering. *Behavioral and Brain Sciences, 30*, 454–455.

Chandrasekharan, S., & Osbeck, L. (2010). Rethinking situatedness: Environmental structure in the time of common code. *Theory and Psychology, 20*(2), 171–207.

Christensen, B. T., & Schunn, C. D. (2008). The role and impact of mental simulation in design. *Applied Cognitive Psychology, 22*, 1–18.

Clancey, W. J. (1997). *Situated cognition: on human knowledge and computer representations*. Cambridge, UK: Cambridge University Press.

Clark, A. (2003). *Natural-born cyborgs: Minds, technologies, and the future of human intelligence*. Oxford, UK: Oxford University Press.

Clark, S., & Corcoran, M. (1986). Perspective on the professional socialization of women: A case of accumulative disadvantage. *Journal of Higher Education, 57*(1), 20–43.

Clewell, B., & Campbell, P. (2002). Taking stock: Where we've been, where we are, where we are going. *Journal of Women and Minorities in Science and Engineering, 8*, 255–284.

Conefrey, T. (1997). Gender, culture and authority in a university life sciences laboratory. *Discourse and Society, 8*(3), 313–340.

Connell, R. (2002). *Gender*. Cambridge: Polity Press.

Craik, K. (1943). *The nature of explanation*. Cambridge: Cambridge University Press.

Damasio, A. R. (1994). *Descartes' error: Emotion, reason, and the human brain*. New York: Grosset/Putnam.

Damasio, A. R. (1996). The somatic marker hypothesis and the possible functions of the prefrontal cortex. *Philosophical Transactions of the Royal Society of London B, 351*, 1413–1420.

Damasio, A. R. (1999). *The feeling of what happens: Body and emotion in the making of consciousness*. New York: Harcourt, Brace.

D'Andrade, R. G. (1995). *The development of cognitive anthropology*. Cambridge: Cambridge University Press.

Danziger, K. (1990). *Constructing the subject: Historical origins of psychological research.* New York: Cambridge University Press.

Danziger, K. (1997). *Naming the mind: How psychology found its language.* London: Sage.

Daston, L., & Galison, P. (2007). *Objectivity.* New York: Zone Books.

Davidson, R. J., & Sutton, S. K. (1995). Affective neuroscience: The emergence of a discipline. *Current Opinion in Neurobiology, 5,* 217–224.

Davies, B., & Harré, R. (1999). Positioning and personhood. In R. Harré & van Langenhove (Eds.), *Positioning theory* (pp. 32–52). Oxford: Blackwell.

Deaux, K., Reid, A., Mizrahi, K., & Ethier, K. A. (1995). Parameters of social identity. *Journal of Personality and Social Psychology, 68,* 280–291.

deFina, A., Schiffrin, D., & Bamburg, M. (Eds.). (2006). *Discourse and identity.* Studies in Interactional Sociolinguistics. New York: Cambridge University Press.

Deigh, J. (1994). Cognitivism in the theory of emotions. *Ethics, 104*(4), 824–854.

De Mey, M. (1982). *The cognitive paradigm.* Chicago: University of Chicago Press.

Denzin, N. K. (1971). The logic of naturalistic inquiry. *Social Forces, 50,* 166–182.

de Sousa, R. (1995). Emotion. In S. Guttenplan (Ed.), *A companion to the philosophy of mind* (p. 270). Oxford: Blackwell.

de Sousa, R. (1997). *The rationality of emotion* (5th printing). Cambridge, MA: MIT Press.

Devos, T., & Banaji, M. (2003). Implicit self and identity. In M. R. Leary & J. B. Tangney (Eds.), *Handbook of self and identity* (pp. 133–175). New York: Guilford.

De Welde, K., Laursen, S., & Thiry, H. (2007). *Women in science, technology, engineering and math (STEM).* Retrieved from http://www.socwomen.org/socactivism/stem_fact_sheet.pdf.

Dewey, J. (1896). The reflex arc concept in psychology. *Psychological Review, 3,* 357–370.

Dewey, J. (1910). *How we think.* New York: Heath.

Dewey, J. (1926). *Art and education.* Merion: Barnes Foundation Press.

Dewey, J. (1929). *Experience and nature.* London: George Allen & Unwin.

Dewey, J. (1930). *The quest for certainty: A study of the relation of knowledge and action.* London: Allen & Unwin.

Dewey, J. (1938a). *Experience and education.* New York: Collier.

Dewey, J. (1938b). *Logic: The theory of inquiry.* New York: Henry Holt.

Dewey, J., & Bentley, A. (1949). *Knowing and the known.* Boston: Beacon.

Dixon, J. (2007). Prejudice, conflict, and conflict resolution. In W. Holloway, H. Lucey, & A. Phoenix (Eds.), *Social psychology matters* (pp. 145–172). Maidenhead, UK: Open University Press.

Donald, M. (1991). *Origins of the Modern Mind: Three Stages in the Evolution of Culture and Cognition.* Cambridge, MA: Harvard University Press.

Dreyfus, H. (1997). Intuitive, deliberative, and calculative models of expert performance. In C. Zsambok & G. Klein (Eds.), *Naturalistic decision making* (pp. 17–28). Mahwah, NJ: Erlbaum.

Dreyfus, H., & Dreyfus, S. (1979). *The scope, limits, and training implications of three models of aircraft pilot emergency response behavior.* Berkeley, CA: Operations Research Center.

Dreyfus, H., & Dreyfus, S. (1980). *A five-stage model of the mental activities involved in directed skill acquisition.* Berkeley, CA: Operations Research Center.

Dunbar, K., & Blanchette, I. (2001). The in vivo/in vitro approach to cognition: The case of analogy. *TRENDS in Cognitive Science, 5,* 334–339.

Duncombe, R. L. (1945). Personal equation in astronomy. *Popular Astronomy, 4*(42), 63-75.

Eberhardt, J. (2005). Imaging race. *American Psychologist, 60*(2), 181–190.

Edwards, D. (1997). *Discourse and cognition.* London: Sage.

Eiduson, B. (1962). *Scientists: Their psychological world.* New York: Basic Books.

Eisenhart, M., & Finkel, E. (2001). Women (still) need not apply. In M. Lederman & I. Bartsch (Eds.), *The gender and science reader* (pp. 13–23). New York: Routledge.

Elgin, C. Z. (1996). *Considered judgment.* Princeton: Princeton University Press.

Ellis, A. (1977). *Anger: How to live with and without it.* Secaucus, NJ: Citadel Press.

Elster, J. (1996). Rationality and the emotions. *Economic Journal, 106,* 1386–1397.

Emirbayer, M., & Goldberg, C. A. (2005). Pragmatism, Bourdieu, and collective emotions in contentious politics. *Theory and Society, 34*(5/6), 469–518.

Endedy, N., Goldberg, J., & Welsh, K. (2005). Complex dilemmas of identity. *Science Education, 90,* 68–90.

Engeström, Y. (1999). Activity theory and individual and social transformation. In Y. Engeström & R. Miettinen (Eds.), *Perspectives on activity theory* (pp. 19–38). Cambridge: Cambridge University Press.

Engeström, Y., & Miettinen, R. (1999). *Perspectives on activity theory.* Cambridge, UK: Cambridge University Press.

Erikson, E. (1959). *Identity and the life cycle.* New York: International Universities Press.

Etzkowitz, H., Kemelgor, C., Neuschatz, M., Uzzi, B., & Alonzo, J. (1994). The paradox of critical mass for women in science. *Science, 266,* 51–54.

Etzkowitz, H., Kemelgor, C., & Uzzi, B. (2000). *Athena unbound: The advancement of women in science.* Cambridge, MA: Harvard University Press.

Evnine, S. (2008). *Epistemic dimensions of personhood.* New York: Oxford University Press.

Fairbairn, W. R. D. (1954). *An object-relations theory of the personality.* New York: Basic Books.

Fanon, F. (1967). *Black skin, white masks.* New York: Grove Press. (Original work published 1952)

Fausto-Sterling, A. (2000). *Sexing the body.* New York: Basic Books.

Feist, G. J. (1998). A meta-analysis of personality in scientific and artistic creativity. *Personality and Social Psychology Review, 2,* 290–309.

Feist, G. J. (2006a). *The psychology of science and the origins of the scientific mind.* New Haven, CT: Yale University Press.

Feist, G. (2006b). The past and future of the psychology of science. *Review of General Psychology, 10*(2), 92–97.

Feist, G. J., & Gorman, M. E. (1998). Psychology of science: Review and integration of a nascent discipline. *Review of General Psychology, 2,* 3–47.

Ferreira, M. (2002). The research lab: A chilly place for graduate women. *Journal of Women and Minorities in Science and Engineering, 8*(1), 85–98.

Ferriman, K., Lubinski, D., & Benbow, C. (2009). Work preferences, life values, and personal views of top math/science graduate students and the profoundly gifted: Developmental changes and gender differences during emerging adulthood and parenthood. *Journal of Personality and Social Psychology, 97*(3), 517–532.

Fetzer, J. (2003). The sky is not falling . . . accepting failure to create innovation in experimentation. *Annals of Biological Chemistry, 375*(5), 597–598.

Feyerabend, P. (1975). *Against method.* Scranton, PA: Verso Books.

Fine, M., Weis, L., Pruitt, L., & Burns, A. (2004). *Off white: Readings on race, power, and society* (2nd ed.). New York: Routledge.

Fink, B. (2007). *Fundamentals of psychoanalytic technique: A Lacanian approach for practitioners.* New York: Norton.

Fodor, J. A. (1980). Methodological solipsism considered as a research strategy in cognitive psychology. *Behavioral and Brain Sciences, 3,* 63–109.

Foucault, M. (1977). What is an author? In D. Bouchard (Ed.), *Language, counter-memory, practice: Selected essays and interviews* (pp. 113–138). Ithaca, NY: Cornell University Press.

Fox, M. F. (2000). Organizational environments and doctoral degrees in science and engineering departments. *Women's Studies Quarterly, 28,* 47–61.

Fox, M. F., & Mohapatra, S. (2007). Social-organizational characteristics of work and publication productivity among academic scientists in doctoral granting departments. *Journal of Higher Education, 78,* 542–571.

Franken, R. (2006). *Human motivation* (6th ed.). Florence, KY: Wadsworth.

Fransella, F. (1995). *George Kelly.* London: Sage.

Freud, S. (1960). *The ego and the id* (G. Strachey, Trans.). New York: Norton. (Original work published 1923)

Fridlund, A. J. (1994). *Human facial expression: An evolutionary view.* San Diego: Academic Press.

Fridlund, A. J., & Russell, J. A. (2006). The functions of facial expressions: What's in a face? In V. Manusov & M. L. Patterson (Eds.), *Sage handbook of nonverbal communication* (pp. 299–320). Newbury Park, CA: Sage.

Galison, P. (1997). *Image and logic: A material culture of microphysics.* Chicago: University of Chicago Press.

Garfinkel, H. (1967). *Studies in ethnomethodology.* Englewood Cliffs, NJ: Prentice-Hall.

Garfinkel, H., & Sacks, H. (1970). On formal structures of practical action. In J. C. McKinney & E. A. Tiryakian (Eds.), *Theoretical sociology: Perspectives and developments* (pp. 338–366). New York: Appleton-Century-Crofts.

Gavey, N. (1989). Feminist poststructuralism and discourse analysis. *Psychology of Women Quarterly, 13,* 459–475.

Gee, J. P. (1999). *An introduction to discourse analysis: Theory and method.* New York: Routledge.

Gee, J. P. (2001). Identity as an analytic lens for research in education. *Review of Research in Education, 25,* 99–125.

Gentner, D. & Stevens, A. L. (1983). *Mental Models.* Hillsdale, NJ: Lawrence Erlbaum Associates.

Gergen, K. J. (2000). *An invitation to social construction.* London: Sage.

Gergen, K. (2001). Self-narration in social life. In M. Wetherell, S. Yates, & S. Taylor (Eds.), *Discourse theory and practice: A reader* (pp. 247–260). London: Sage.

Gibson, J. J. (1979). *The ecological approach to visual perception*. Boston: Houghton-Mifflin.

Giere, R. N. (1988). *Explaining science: A cognitive approach*. Chicago: University of Chicago Press.

Giere, R. N. (Ed.). (1992). *Cognitive models of science*. Minnesota Studies in the Philosophy of Science, Vol. 15. Minneapolis: University of Minnesota Press.

Gilbert, N., & Mulkey, M. (1984). *Opening Pandora's box*. Cambridge: Cambridge University Press.

Giorgi, A. (1970). *Psychology as a human science: A phenomenologically based approach*. New York: Harper & Row.

Glaser, B. G. (1992). Basics of grounded theory analysis: *Emergence vs forcing*. Mill Valley, CA: Sociology Press.

Glaser, B., & Strauss, A. (1967). *The discovery of grounded theory: Strategies for qualitative research*. Piscataway, NJ: Aldine Transaction.

Godfrey-Smith, P. (2002). Dewey on naturalism, realism and science. *Philosophy of Science, 69*, S25–S35.

Goffman, E. (1981). *Forms of talk*. Philadelphia: University of Pennsylvania Press.

Gooding, D. C. (1985). In nature's school: Faraday as a natural philosopher. In D. Gooding & F. James (Eds.), *Faraday rediscovered* (pp. 105–135). London: Macmillan.

Gopnik, A. (1996). The scientist as a child. *Philosophy of Science, 63*, 485–514.

Gorman, M. (1997). Mind in the world: Cognition and practice in the invention of the telephone. *Social Studies of Science, 27*, 583–624.

Gorman, M. E., & Carlson, W. B. (1990). Interpreting invention as a cognitive process: The case of Alexander Graham Bell, Thomas Edison, and the telephone. *Science, Technology, and Human Values, 15*, 131–164.

Gorman, M. E., Tweney, R. D., Gooding, D., & Kincannon, A. P. (2005). *Scientific and technological thinking*. Mahwah, NJ: Erlbaum.

Graver, M. (2007). *Stoicism and emotion*. Chicago: University of Chicago Press.

Greeno, J. G. (1989). Situations, mental models, and generative knowledge. In D. Klahr & K. Kotovsky (Eds.), *Complex information processing* (pp. 285–318). Hillsdale, NJ: Erlbaum.

Greeno, J. G. (1998). The situativity of knowing, learning, and research. *American Psychologist, 53*, 5–26.

Greenspan, P. (2003). Emotions, rationality, and mind/body. In A. Hatzimoysis (Ed.), *Philosophy and the emotions* (pp. 113–126). Cambridge: Cambridge University Press.

Grene, M. (2002). Intellectual autobiography. In R. Auxier & L. E. Hahn (Eds.), *The philosophy of Marjorie Grene* (pp. 3–28). Library of Living Philosophers Series. Peru, IL: Open Court.

Griffiths, P., & Scarantino, A. (2008). Emotions in the wild: The situated perspective on emotion. In P. Robbins & M. Aydede (Eds.), *The Cambridge handbook of situated cognition* (pp. 437–453). Cambridge: Cambridge University Press.

Gross, D. M. (2006). *The secret history of emotion*. Chicago: University of Chicago Press.

Guthrie, R. V. (1998). *Even the rat was white*. Boston, MA: Allyn and Bacon.

Haidt, J. (2001). The emotional dog and its rational tail: A social intuitionist approach to moral judgment. *Psychological Review, 108*, 814–834.

Haidt, J. (2007). The new synthesis in moral psychology. *Science, 18*(316), 998–1002.

Hall, R., Stevens, R., & Torralba, T. (2002). Disrupting representational infrastructure in conversation across disciplines. *Mind, Culture, and Activity, 9*, 179–210.

Hall, S. (2001). Foucault: Knowledge and discourses. In M. Wetherell, S. Taylor, & S. Yates (Eds.), *Discourse theory and practice: A reader* (pp. 72–81). London: Sage.

Hall, R., Wickert, K., & Wright, K. (2010). How does cognition get distributed? Case studies of making concepts general in technical and scientific work. In M. Banich & D. Caccamise (Eds.), *Generalization of knowledge: Multidisciplinary perspectives*. New York: Psychology Press.

Hamilton, K. (2004). Faculty science positions continue to elude women of color. *Black Issues in Higher Education, 21*(23), 36–40.

Haraway, D. (1991). *Simians, cyborgs, and women*. New York: Routledge.

Harding, S. (1986). *The science question in feminism*. Ithaca, NY: Cornell University Press.

Harding, S. (1996). Rethinking standpoint epistemology: What is 'strong objectivity'? In E. F. Keller & H. E. Longino (Eds.), *Feminism and science* (pp. 235–248). Oxford: Oxford University Press.

Harmon, E., & Nersessian, N. J. (2008). Cognitive partnerships on the benchtop: Designing to support scientific researchers. In *Proceedings of the 7th ACM Conference on Designing Interactive Systems – DIS 2008* (pp. 119–128). New York: ACM.

Harré, R. (1986). *The social construction of emotions*. London: Blackwell.

Harré, R. (1992). What is real in psychology: A plea for persons. *Theory and Psychology, 2*(2), 153–158.

Harré, R. (1998). *The singular self: An introduction to the psychology of personhood*. London: Sage.

Harré, R., & Gillett, G. (1994). *The discursive mind*. London: Sage.

Harré, R., & Moghaddam, F. (2003). Introduction: The self and others in traditional psychology and in positioning theory. In R. Harré & F. Moghaddam (Eds.), *The self and others: Positioning individuals and groups in personal, political, and cultural contexts* (pp. 1–11). Westport, PA: Praeger.

Harré, R., & Moghaddam, F. M., Cairnie, T. P., Rothbart, D., & Sabat, S. R. (2009). Recent advances in positioning theory. *Theory and Psychology, 19*(1), 5–31.

Harré, R., & Van Langenhove, L. (1999). The dynamics of social episodes. In R. Harré & L. Van Langenhove (Eds.), *Positioning theory: moral contexts of intentional action* (pp. 1–13). Oxford, Blackwell.

Heberlein, A., & Adolphs, R. (2004). Impaired spontaneous anthropomorphizing despite intact perception and social knowledge. *Proceedings of the National Academy of Sciences, 101*(19), 7487–7491.

Heidegger, M. (1962). *Being and time* (J. Macquarrie & E. Robinson, Trans.). New York: Harper and Row.

Held, B. S. (2007). *Psychology's interpretive turn: The search for truth and agency in theoretical and philosophical psychology*. Washington, DC: American Psychological Association.

Hempel, C. (1952). *Fundamentals of concept formation in empirical science*. Chicago: University of Chicago Press.

Henwood, F., & Miller, K. (2001). Boxed in or coming out? On the treatment of science, technology and gender in educational research. *Gender and Education, 13*, 237–242.

Hesse, M. B. (1966). *The explanatory function of metaphor*. Notre Dame, IN: University of Notre Dame Press.

Holland, J. L. (1973). *Making vocational choices: A theory of careers*. Englewood Cliffs, NJ: Prentice-Hall

Horkheimer, M., & Adorno, T. (1973). *The dialectics of enlightenment*. New York: Continuum. (Original work published 1944)

Horney, K (1945). *Our inner conflicts: A constructive theory of neurosis*. New York: Norton.

Horowitz, A., & Bekoff, M. (2007). Naturalizing anthropomorphism: Behavioral prompts to our humanizing of animals. *Anthrozoös, 20*(1), 23–35.

Huffman, K. (2010). *Psychology in action* (9th ed.). New York: John Wiley & Sons.

Hunter, A. B., Laursen, S. L., & Seymour, E. (2006). Becoming a scientist: The role of undergraduate research in students' cognitive, personal, and professional development. *Science Education, 91*(1), 36–74.

Hutchins, E. (1995a). How a cockpit remembers its speeds. *Cognitive Science, 19*, 265–288.

Hutchins, E. (1995b). *Cognition in the wild*. Cambridge, MA: MIT Press.

Isaacson, W. (2007). *Einstein: His life and universe*. New York: Simon & Schuster.

James, W. (1890). *The principles of psychology* (Vols. 1–2). New York: Henry Holt.

James, W. (1884). What is an emotion? *Mind, 9*, 188–205.

James, W. (1896). *The sentiment of rationality*. (Reprinted from *The will to believe and other essays*). Cambridge, MA: University Press.

James, W. (2003). *Pragmatism: A new name for some old ways of thinking*. New York: Barnes and Noble. (Original work published 1907)

Jenkins, A. H. (1995). *Psychology of African Americans: A humanistic approach*. Needham, MA: Allyn & Bacon.

Johnson, A. (2007). Unintended consequences: How science professors discourage women of color. *Science Education, 91*, 805–821.

Johnson, M. (2007). *The meaning of the body*. Chicago: University of Chicago Press.

Johnson-Bailey, J. (2002). Race matters: The unspoken variable in the teaching-learning transaction. *New Directions for Adult and Continuing Education, 93*, 39–49.

Johnson-Bailey, J. (2004). Hitting and climbing the proverbial wall: Participation and retention issues for black graduate women. *Race, Ethnicity and Education, 7*, 231–249.

Johnson-Laird, P. N. (1983). *Mental Models*. Cambridge, MA: Harvard University Press.

Jones, C., & Shorter-Gooden, K. (2003). *Shifting: The double lives of black women in America*. New York: Harper Collins.

Jordan, B., & Henderson, A. (1995). Interaction analysis: Foundations and practice. *Journal of the Learning Sciences, 4*(1), 39–103.

Jordan, D. (2006). *Sisters in science: Conversations with black women scientists about race, gender, and their passion for science.* West Lafayette, IN: Purdue University Press.

Kant, I. (1965). Section 16 of section of *Analytic of concepts.* (Reprinted from N. K. Smith, Trans., *Transcendental logic,* London: R & R. Clark).

Keller, E. F. (1995). *Reflections on gender and science.* New Haven, CT: Yale University Press. (Original work published 1985)

Keller, E. F. (1996). Language and ideology in evolutionary theory: Reading cultural norms into natural law. In E. F. Keller & H. E. Longino (Eds.) (pp. 154–172). *Feminism and science.* Oxford: Oxford University Press.

Keller, E. F. (2001). Gender and science: An update. In M. Wyer, M. Barbercheck, D. Geisman, H. Öztürk, & M. Wayne (Eds.), *Women, science, and technology: A reader in feminist science studies* (pp. 132–140). New York: Routledge.

Keller, E. F. & Grontkowski, C. (1996). The mind's eye. In E. F. Keller & H. E. Longino (Eds.), *Feminism and science* (pp. 187–202). Oxford: Oxford University Press. (Originally published 1983)

Keller, E. F., & Longino, H. E. (1996). *Feminism and science.* Oxford, UK: University Press.

Kelly, G. A. (1955). *The psychology of personal constructs* (Vols. 1–2). New York: Norton.

Kelly, G. A. (1963). *A theory of personality: The psychology of personal constructs.* New York: Norton.

Kelly, G. A. (1964). The language of hypothesis. *Journal of Individual Psychology, 20,* 137–154.

Kelly, G. (1969). Man's construction of his alternatives. In B. Maher (Ed.), *Clinical psychology and personality: The selected papers of George Kelly* (pp. 66–93). New York: Wiley.

Kerr, E. A. (2001). Toward a feminist natural science: Linking theory and practice. In M. Lederman & I. Bartsch (Eds.), *The gender and science reader* (pp. 386–406). New York: Routledge.

Kiesling, S. (2006). Hegemonic identity making. In A. deFina, D. Schiffrin, & M. Bamburg (Eds.), *Discourse and identity* (pp. 261–287). Studies in Interactional Sociolinguistics. New York: Cambridge University Press.

Kinder, D. R., & Sears, D. O. (1981). Prejudice and politics: Symbolic racism versus racial threats to the good life. *Journal of Personality and Social Psychology, 40*(3), 414–431.

Kirschner, S. (2006). Psychology and pluralism: Toward the psychological studies. *Journal of Theoretical and Philosophical Psychology, 26*(1–2), 1–17.

Kitcher, P. (1993). *The advancement of science.* Oxford: Oxford University Press.

Klahr, D. (2000). *Exploring science: The cognition and development of discovery processes.* Cambridge, MA: MIT Press.

Klahr, D., & Simon, H. A. (1999). Studies of scientific discovery: Complementary approaches and convergent findings. *Psychological Bulletin, 125,* 524–543.

Knorr Cetina, K. (1999). *Epistemic cultures: How the sciences make knowledge.* Cambridge, MA: Harvard University Press.

Koch, S. (1992). Wundt's creature at age zero – and as centurian: Some aspects of the institutionalization of psychology. In S. Koch & D. Leary (Eds.), *A century of psychology as science.* Washington, DC: American Psychological Association.

Koch, S. (1993). "Psychology" or "the psychological studies"? *American Psychologist,* *48*(8), 902–904.

Konstan, D. (2006). *The emotions of the ancient Greeks: Studies in Aristotle and Greek literature.* Toronto: University of Toronto Press.

Korobov, N. (2010). A discursive psychological approach to positioning. *Qualitative Research in Psychology.*

Kövecses, Z. (2000). *Metaphor and emotion. Language, culture and body in human feeling.* Cambridge: Cambridge University Press.

Kruglanski, A., & Ajzen, I. (1983). Bias and error in human judgment. *European Journal of Social Psychology, 19,* 448–468.

Kuhn, T. (1962). *The structure of scientific revolutions.* Chicago: University of Chicago Press.

Kuhn, T. (1977). Objectivity, value judgment, and theory choice. In T. Kuhn (Ed.), *The essential tension: Selected studies in scientific tradition and change* (pp. 320–339). Chicago: University of Chicago Press.

Kurz-Milcke, E., Nersessian, N. J., & Newstetter, W. (2004). What has history to do with cognition? Interactive methods for studying research laboratories. *Journal of Cognition and Culture, 4,* 663–700.

Lacan, J. (1998). *The seminar of Jacques Lacan. Book XX: Encore, 1972–1973* (J.-A. Miller, Ed., & B. Fink, Trans.). New York: Norton.

Lacey, H. (1999). *Is science value free? Values and scientific understanding.* London: Routledge.

Lakoff, G. (1987). *Women, fire, and dangerous things: What categories reveal about the mind.* Chicago: University of Chicago Press.

Lakoff, G., & Johnson, M. (1980). *Metaphors we live by.* Chicago: University of Chicago Press.

Lakoff, G., & Johnson, M. (1999). *Philosophy in the flesh.* New York: Basic Books.

Lakoff, G., & Nunez, R. (2000). *Where mathematics comes from: How the embodied mind brings mathematics into being.* New York: Basic Books.

Lamiell, J. (2003). *Beyond individual and group differences: Human individuality, scientific psychology, and William Stern's critical personalism.* Thousand Oaks, CA: Sage.

Lamiell, J. T. (2009). Some philosophical and historical considerations relevant to William Stern's contributions to developmental psychology. *Zeitschrift für Psychologie, 217*(2), 66–72.

Lamiell, J. (2010). Translation of William Stern's "Psychology and personalism." *New Ideas in Psychology, 28,* 110–134.

Laqueur, T. (1990). *Making sex: Body and gender from the Greeks to Freud.* Cambridge, MA: Harvard University Press.

Latour, B. (1987). *Science in action.* Cambridge, MA: Harvard University Press.

Latour, B. (2005). *Reassembling the social. An introduction to actor network theory.* New York: Oxford University Press.

Latour, B., & Woolgar, S. (1986). *Laboratory life: The construction of scientific facts.* Princeton, NJ: Princeton University Press. (Original work published 1979)

Lave, J. (1988). *Cognition in practice.* Cambridge: Cambridge University Press.

Leggon, C. (1997). The scientist as academic. *Daedalus, 126*(4), 221–245.

Lave, J. (1988). *Cognition in practice: Mind, mathematics, and culture in everyday life.* New York: Cambridge University Press.

Lave, J., & Wegner, E. (1991). *Situated learning: Legitimate peripheral participation.* New York: Cambridge University Press.

Lazarus, R. S. (2006). *Stress and emotion: A new synthesis.* New York: Springer.

Lazarus, R. S., & Folkman, S. (1987). Transactional theory and research on emotions and coping. *European Journal of Personality, 1,* 141–169.

Lazarus, A., Averill, J. R., & Opton, E. M. (1970). Toward a cognitive theory of emotions. In M. B. Arnold (Ed.), *Feelings and emotions* (pp. 207–232). New York: Academic Press.

LeDoux, J. (1996). *The Emotional brain: The mysterious underpinnings of emotional life,* New York: Simon and Schuster.

LeDoux, J. (1998). Fear and the brain: Where have we been, and where are we going? *Biological Psychiatry, 44*(12), 1229–1238.

Leont'ev, A. N. (1978). *Activity, consciousness, and personality.* Englewood Cliffs, NJ: Prentice-Hall.

Lewins, R., & Lewontin, R. (1985). *The dialectical biologist.* Cambridge, MA: Harvard University Press.

Lloyd, G. (1996). Reason, science, and the domination of matter. In E. F. Keller & H. E. Longino (Eds.), *Feminism and science* (pp. 41–56). Oxford: Oxford University Press. (Original work published 1993)

Longino, H. (1990). *Science as social knowledge: Values and objectivity in scientific inquiry.* Princeton, NJ: Princeton University Press.

Longino, H. E. (2001). Can there be a feminist science? In M. Wyer, M. Barbercheck, D. Geisman, H. Öztürk, & M. Wayne (Eds.), *Women, science, and technology: A reader in feminist science studies* (pp. 216–222). New York: Routledge.

Longino, H. (2002). *The fate of knowledge.* Princeton, NJ: Princeton University Press.

Louis, W. R. (2008). Intergroup positioning and power. In F. M. Moghaddam, R. Harré, & N. Lee (Eds.), *Global conflict resolution through positioning analysis* (pp. 21–39). New York: Springer.

Lutz, C., & White, G. M. (1986). The anthropology of emotions. *Annual Review of Anthropology, 15,* 405–436.

Lynch, M. (1993). *Scientific practice and ordinary action: Ethnomethodology and social studies of science.* Cambridge: Cambridge University Press.

Machamer, P., & Osbeck, L. (2004). The social in the epistemic. In P. Machamer & G. Wolters (Eds.), *Science, values, and objectivity.* Pittsburgh: University of Pittsburgh Press.

Mahoney, M. (2004). *Scientist as subject: The psychological imperative.* Cambridge, MA: Ballinger Publishing. (Original work published 1976)

Malone K. R., & Barabino, G. (2009). Narrations of race in STEM settings: Identity formation and its discontents. *Science Education, 93*(3), 485–510.

Malone, K. R., & Kelly, S. D. (2006). Women in science: Should we bother with a psychoanalytic viewpoint? *Psychoanalysis, Culture & Society, 10,* 207.

Malone, K., Nersessian, N. J., & Newstetter, W. (2005). Gender writ small: Gender enactments in organization and knowledge transmission in a biomedical engineering laboratory. *Journal of Women and Minorities in Science and Engineering, 11,* 61–82.

Mannix, M. (2002). Facing the problem, *American Society for Engineering Education Prism, 12*(2), 19–24.

Manuel, F. E. (1968). *A portrait of Isaac Newton.* Cambridge, MA: Harvard University Press.

Maquet, J. (1993). Objects as instruments, objects as signs. In S. Lubar & W. D. Kingery (Eds.), *History from things: Essays on material culture* (pp. 30–40). Washington, DC: Smithsonian Institution.

Margolis, J. (2003). *The unraveling of scientism.* Ithaca, NY: Cornell University Press.

Martin, J., Sugarman, J. H., & Thompson, J. (2003). *Psychology and the question of agency.* Albany, NY: SUNY Press.

Maslow, A. (1966). *The psychology of science: A reconnaissance.* New York: Harper & Row.

Maslow, A. H. (1987). *Motivation and personality.* New York: Harper & Bros. (Original work published 1954)

Matthews, G., Zeidner, M., & Roberts, R. D. (2002). *Emotional intelligence. Science and myth.* Cambridge, MA: MIT Press.

McAllister, J. W. (1996). *Beauty and revolution in science.* Ithaca, NY: Cornell University Press.

McGuire, T., & Tuchanska, B. (2001). *Science unfettered: Philosophical study in sociohistorical ontology.* Athens, OH: Ohio University Press.

McIlwee, J., & Robinson, J. (1992). *Women in engineering.* Albany, NY: SUNY Press.

Mead, G. H. (1932). *The philosophy of the present.* La Salle, IL: Open Court Publishing.

Mead, G. H. (1934). *Mind, self, and society: From the standpoint of a social behaviorist* (C. W. Morris, Ed.). Chicago: University of Chicago Press.

Menary, R. (2007). *Cognitive integration: Mind and cognition unbounded.* New York: Palgrave MacMillan.

Mills, C. W. (1997). *The racial contract.* Ithaca, NY: Cornell University Press.

Mitroff, I. (1974). *The subjective side of science: Philosophical inquiry into the psychology of the Apollo Moon Scientists* Amsterdam: Elsevier.

Moghaddam, F. (1998). *Social psychology: Exploring universals across cultures.* New York: W. H. Freeman.

Moghaddam, F. (1999). Reflexive positioning: Culture and private discourse. In R. Harré and L. von Langenhove, *Positioning theory* (pp. 74–86). Oxford, UK: Blackwell

Moody, J. (2003). Recruiting and retaining women and minority faculty: An interview with JoAnn Moody (interview by Nancy E. Carriuola). *Journal of Developmental Education, 27,* 18–34.

Moore, S., & Oaksford, M. (2002). *Emotional cognition.* Amsterdam: John Benjamins.

Morrow, S. L. (2005). Quality and trustworthiness in qualitative research in counseling psychology. *Journal of Counseling Psychology, 52,* 250–260.

Mulkay, M., & Gilbert, G. N. (1982). Accounting for error: How scientists construct their social world when they account for correct and incorrect belief. *Sociology, 16*(2), 165–183.

Myers, D. (2002). Discrimination. In *Social psychology* (7th ed., p. 330). Boston: McGraw Hill(6).

Nasir, S., & Saxe, G. (2003). Ethnic and academic identities: A cultural practice perspective on emerging tensions and their management in the lives of minority students. *Educational Researcher, 32*(5), 14–18.

Neal, M. A. (2005). *New black man.* New York: Routledge.

Nelson, D. J., & Rogers, D. C. (2002). *A national analysis of diversity in science and engineering faculties at research universities.* Retrieved from http://www.now.org/issues/diverse/diversity_report.pdf.

Nelson, L. H. (1990). *Who knows: From Quine to a feminist empiricism.* Philadelphia: Temple University Press.

Nersessian, N. J. (1984). *Faraday to Einstein: Constructing meaning in scientific theories.* Dordrecht: Martinus Nijhoff/Kluwer.

Nersessian, N. J. (1992a). How do scientists think? Capturing the dynamics of conceptual change in science. In R. N. Giere (Ed.), *Cognitive models of science* (pp. 3–45). Minneapolis: University of Minnesota Press.

Nersessian, N. J. (1992b). In the theoretician's laboratory: Thought experimenting as mental modeling. *PSA 1992, 2,* 291–301.

Nersessian, N. J. (1995a). Opening the black box: Cognitive science and the history of science. *Osiris, 10,* 194–211.

Nersessian, N. J. (1995b). Should physicists preach what they practice? Constructive modeling in doing and learning physics. *Science & Education, 4*(3), 203–226.

Nersessian, N. J. (2002). The cognitive basis of model-based reasoning in science. In P. Carruthers, S. Stich, & M. Siegal (Eds.), *The cognitive basis of science* (pp. 133–153). Cambridge: Cambridge University Press.

Nersessian, N. J. (2005). Interpreting scientific and engineering practices: Integrating the cognitive, social, and cultural dimensions. In M. E. Gorman, R. D. Tweney, D. C. Gooding, & A. P. Kincannon (Eds.), *Scientific and technological thinking* (pp. 17–56). Hillsdale, NJ: Erlbaum.

Nersessian, N. J. (2006a). The cognitive-cultural systems of the research laboratory. *Organization Studies, 27*(1), 125–145.

Nersessian, N. J. (2006b). Model-based reasoning in distributed cognitive systems. *Philosophy of Science, 72,* 699–709.

Nersessian, N. J. (2008a). *Creating scientific concepts.* Cambridge, MA: MIT Press.

Nersessian, N. J. (2008b). Mental modeling in conceptual change. In S. Vosniadou (Ed.), *International handbook of conceptual change* (pp. 391–416). New York: Routledge.

Nersessian, N. (2009). How do engineering scientists think? Model-based simulation in biomedical engineering research laboratories. *Topics in Cognitive Science, 1*(3).

Nersessian, N. J., & Chandrasekharan, S. (2009). Hybrid analogies in conceptual innovation in science [Special issue]. *Cognitive Systems Research Journal, 10,* 178–188.

Nersessian, N. J., Kurz-Milcke, E., & Davies, J. (2005). Ubiquitous computing in science and engineering research laboratories: A case study from biomedical engineering. In M. S. Gerassimos Kouzelis, M. Pournari, & V. Tselfes (Eds.), *Knowledge in the new technologies* (pp. 167–195). Berlin: Peter Lang.

Nersessian, N. J., Kurz-Milcke, E., Newstetter, W. C., & Davies, J. (2003). Research laboratories as evolving distributed cognitive systems. *Proceedings of the 25th Annual Conference of the Cognitive Science Society,* 857–862.

Nersessian, N. J., Newstetter, W. C., Kurz-Milcke, E., & Davies, J. (2003). A mixed-method approach to studying distributed cognition in evolving environments. *Proceedings of the International Conference on Learning Sciences,* 307–314.

Nersessian, N. J., & Patton, C. (2009). Model-based reasoning in interdisciplinary engineering. In A. W. M. Meijers (Ed.), *The handbook of the philosophy of technology & engineering sciences* (pp. 678–718). New York: Springer.

Neu, J. (2000). *A tear is an intellectual thing: The meaning of emotions.* Oxford: Oxford University Press.

Newell, Allen; Shaw, J. C., & Simon, Herbert A. (1958). Elements of a theory of human problem solving. *Psychological Review, 65*(3), May, 1958. 151–166.

Newell, A., & Simon, H. A. (1976). Computer science as empirical enquiry. *Communications of the ACM, 19,* 113–126.

Newstetter, W. (2005). Designing Cognitive Apprenticeships in Biomedical Engineering. *Journal of Engineering Education, 94*(2), 207–213.

Newstetter, W., Kurz-Milcke, E., & Nersessian, N. J. (2004). Cognitive partnerships on the bench tops. *Proceedings of 2004 ICLS Conference,* 372–379.

Newstetter, W., Nersessian, N. J., Kurz-Milcke, E., & Malone, K. R. (2002). Laboratory learning, classroom learning: Looking for convergence/divergence in biomedical engineering. In *Proceedings of the International Conference on Learning Sciences* (pp. 315–321). Hillsdale, NJ: Lawrence Erlbaum.

Nosek, B. A., Banaji, M. R., & Greenwald, A. G. (2002). Math = male, me = female, therefore math ≠ me. *Journal of Personality and Social Psychology, 83*(1), 44–59.

Nussbaum, M.C. (2001). *Upheavals of thought: The intelligence of emotions.* Cambridge: Cambridge University Press.

Olsen, D., Maple, S. A., & Stage, F. (1995). Women and minority faculty job satisfaction: Professional role interests, professional satisfactions, and institutional fit. *The Journal of Higher Education, 66*(3), 267–293.

Olson, K. (2002). Who gets promoted? Gender differences in science and engineering academia. *Journal of Women and Minorities in Science and Engineering, 8,* 347–362.

Ortner, S. (1995). Resistance and the problem of ethnographic refusal. *Comparative Studies in Society and History, 37*(1), 173–193.

Osbeck, L., & Good, J. M. M. (2005, August). *Re-presenting representation: "Distributed cognition," innovation, and conceptual change in science.* Paper presented at the Annual Meeting of the American Psychological Association, Washington, DC.

Osbeck, L. M., Malone, K. R., & Nersessian, N. J. (2007). Dissenters in the sanctuary: Evolving frameworks in "mainstream" cognitive science. *Theory and Psychology, 17*(2), 243.

Osbeck, L., & Nersessian, N. J. (2006). The distribution of representation. *Journal for the Theory of Social Behavior, 36*(2), 141–160.

Osbeck, L., Newstetter, W., & Nersessian, N. J. (2006). Positioning in the laboratory. Paper presented at International Society for Psychology of Science. Zacatecas, Mexico.

Papadopoulos, D. (2008). In the ruins of representation: Identity, individuality, subjectification. *British Journal of Social Psychology, 47*(1), 139–165.

Parkinson, B., Fischer, A. H., & Manstead, A. S. R. (2005). *Emotions in social relations: Cultural, group, and interpersonal processes.* New York: Psychology Press.

Parrott, G., & Harré, R. (1996). *The emotions: Social, cultural, and biological dimensions.* London: Sage.

Perrig, W., & Kintsch, W. (1985). Propositional and situational representations of text. *Journal of Memory and Language, 24,* 503–518.

Petroski, H. (2008). *Success through failure: The paradox of design.* Princeton, NJ: Princeton University Press.

Pham, M. T. (2007). Emotion and rationality: A critical review and interpretation of empirical evidence. *Review of General Psychology, 11*(2), 155–178.

Pickering, A. (1984). *Constructing quarks: A sociological history of particle physics.* Chicago: University of Chicago Press.

Pickering, A. (1987). Forms of life: Science, contingency, and Harry Collins. *British Journal for the History of Science, 20,* 213–221.

Pickering, A. (1995). *The mangle of practice: Time, agency, & science.* Chicago: University of Chicago Press.

Poggenpoel, M., & Myburgh, C. (2003). The researcher as research instrument in educational research: A possible threat to trustworthiness? *Education, 124,* 418–421.

Pols, E. (1982). *The acts of our being: a reflection on agency and responsibility.* Amherst: University of Massachusetts Press.

Polanyi, M. (1964). *Science, faith, and society.* Chicago: University of Chicago Press. (Original work published 1946)

Polanyi, M. (1973). *Personal knowledge: Towards a post-critical philosophy.* Chicago: University of Chicago Press. (Original work published 1958)

Prediger, D. J. (1982). Dimensions underlying Holland's hexagon: Missing link between interests and occupations? *Journal of Vocational Behavior, 21*(3), 259–287.

Prinz, J. (2007). *The emotional construction of morals.* Oxford: Oxford University Press.

Proctor, H., & Parry, G. (1978). Constraint and freedom: The social origin of personal constructs. In F. Fransella (Ed.), *Personal construct psychology 1977* (pp. 157–170). London: Academic Press.

Protevi, J. (2007). Series description. *New directions in philosophy and cognitive science.* New York: Palgrave.

Redman, P. (2005). The narrative formation of identity revisited: Narrative construction, agency and the unconscious. *Narrative Inquiry, 15*(1), 25–44.

Reichenbach, H. (1938). *Experience and prediction.* Chicago: University of Chicago Press.

Resnick, L. B. (1996). Situated learning. In E. De Corte & F. E. Weinert (Eds.), *International encyclopedia of developmental and instructional psychology* (pp. 341–347). Oxford: Elsevier Science.

Ribeiro, B. (2006). Footing, positioning, voice: Are we talking about the same thing. In A. deFina, D. Schiffrin, & M. Bamburg (Eds.), *Discourse and identity* (pp. 48–82). Cambridge: Cambridge University Press.

Richards, G. (1996). *Putting psychology in its place.* London: Routledge.

Robinson, D. N. (1989). *Aristotle's psychology.* New York: Columbia University Press.

Robinson, D. N. (2007). Theoretical psychology: What is it and who needs it? *Theory and Psychology, 17*(2), 187–198.

Rosch, E. (1975). Cognitive representations of semantic categories. *Journal of Experimental Psychology: General, 104*(3), 192–233.

Roseman, I. J. (1991). Appraisal determinants of discrete emotions. *Cognition and Emotion, 5*, 161–200.

Rosser, S. (1997). *Re-engineering female friendly science.* New York: Teachers College Press.

Rosser, S. V. (1999). Different laboratory/work climates: Impacts on women in the workplace. *Annals of the New York Academy of Sciences, 869*, 95–101.

Rosser, S. V. (2001). Are there methodologies appropriate for the natural sciences and do they make a difference? In M. Lederman & I. Bartsch (Eds.), *The gender and science reader* (pp. 123–144). New York: Routledge.

Rosser, S. V. (2004). *The science glass ceiling: Academic women scientists and the struggle to succeed.* New York: Routledge.

Rouse, J. (1996). *Engaging science: How to understand its practices philosophically.* Ithaca, NY: Cornell University Press.

Rouse, J. (2002). *How scientific practices matter: Reclaiming philosophical naturalism.* Chicago: University of Chicago Press.

Rowlands, M. (1999). *The body in mind.* Cambridge: Cambridge University Press.

Rychlak, J. F. (1990). George Kelly and the concept of construction. *International Journal of Personal Construct Psychology, 3*(1), 7–20.

Rychlak, J. F. (1997). *In defense of human consciousness.* Washington, DC: American Psychological Association.

Ryle, G. (1949). *The concept of mind.* Chicago: University of Chicago Press.

Schaffer, S. (1988). Astronomers mark time: Discipline and the personal equation. *Science in Context, 2*, 115–145.

Schatzki, T., Knorr Cetina, K., & von Savigny, E. (Eds.). (2001). *The practice turn in contemporary theory.* London: Routledge

Schiebinger, L. (1999). *Has feminism changed science?* Cambridge, MA: Harvard University Press.

Schiebinger, L. (2001). Creating sustainable science. In M. Lederman & I. Bartsch (Eds.), *The gender and science reader* (pp. 466–483). New York: Routledge.

Schiebinger, L. (2002). Mainstreaming gender analysis into science. *Journal of Women and Minorities in Science and Engineering, 8*, 381–394.

Scott, J. (1996). Gender: A useful category of historical analysis. In J. Scott (Ed.), *Feminism and history* (pp. 152–182). Oxford: Oxford University Press.

Seymour, E., & Hewitt, N. (1997). *Talking about leaving: Why undergraduates leave the sciences.* Boulder, CO: Westview Press.

Seymour, E., Hunter, A. B., Laursen, S. L., & Deantoni, T. (2004). Establishing the benefits of research experiences for undergraduates in the sciences: First findings from a three year study. *Science Education, 88*(4), 493–534.

Shepardson, C. (1998). Human diversity and the sexual relation. In C. Lane (Ed.), *Psychoanalysis and race* (pp. 41–64). New York: Columbia University Press

Shoemaker, S. (1984). Personal identity: A materialist's account. In S. Shoemaker & R. Swinburne (Eds.), *Personal identity* (pp. 89–97). Oxford: Blackwell.

Shore, B. (1996). *Culture in mind: Cognition, culture, and the problem of meaning.* Oxford: Oxford University Press.

Simon, H. A., Langley, P. W., & Bradshaw, G. (1981). Scientific discovery as problem solving. *Synthese, 47*, 1–27.

Simonton, D. K. (1988). *Scientific genius: A psychology of science.* Cambridge: Cambridge University Press.

Simonton, D. K. (2004). *Creativity in science.* Cambridge: Cambridge University Press.

Sinclair, B. (2004). *Technology and the African-American experience.* Cambridge, MA: MIT Press.

Skinner, B. F. (1938). *The behavior of organisms: An experimental analysis.* New York: Appleton-Century-Crofts.

Skinner, B. F. (1953). *Science and human behavior.* New York: Macmillan.

Smith J. A. (Ed.) (2003). *Qualitative psychology: A practical guide to research methods.* London: Sage.

Smythe, W. (1998). *Toward a psychology of persons.* Mahwah, NJ: Erlbaum.

Solomon, R. C. (1976). *The passions.* New York: Doubleday.

Solomon, R. C. (1993). The philosophy of emotions. In M. Lewis & J. M. Haviland (Eds.), *Handbook of emotions* (pp. 3–15). New York: Guilford.

Solomon, R. (2003). *Not passion's slave: Emotions and choice.* New York: Oxford University Press.

Sonnert, G., & Holton, G. (1995). *Who succeeds in science?* New Brunswick, NJ: Rutgers University Press.

Spradley, J. (1979). *The ethnographic interview.* New York: Holt, Rinehart, and Winston.

Steele, C. M. (2003). Through the back door to theory. *Psychological Inquiry, 14,* 314–317.

Steele, J. (2003). Children's gender stereotypes about math: The role of stereotype stratification. *Journal of Applied Social Psychology, 33*(12), 2587–2606.

Stepan, N. L. (1996). Race and gender: The role of analogy in science. In E. F. Keller & H. Longino (Eds.), *Feminism and science* (pp. 121–136). Oxford: Oxford University Press.

Stern, W. (1906). *Person und Sache: System der philosophischen Weltanschauung: Band l. Ableitung und Grundlehre* [Person and world: A system of a philosophical worldview: Vol. 1. Introduction and basic principles]. Leipzig: Barth.

Stem, W. (1917). *Die Psychologie und der Personalismus* [Psychology and personalism]. Leipzig: Barth.

Sternberg, R. J. (2001). *Psychology: In search of the human mind* (3rd ed.). Ft. Worth, TX: Harcourt College Publishers.

Storer, N. W. (1966). *The social system of science.* New York: Holt, Rinehart, & Winston.

Strauss, A. (1987). *Qualitative analysis for social scientists.* Cambridge: Cambridge University Press

Strauss, A., & Corbin, J. (1998). *Basics of qualitative research techniques and procedures for developing grounded theory* (2nd ed.). London: Sage.

Stryker, S., & Burke, P. (2000). The past, present and future of identity theory. *Social Psychology Quarterly, 63,* 284–297.

Suchman, L. (2007). *Human-machine reconfigurations: Plans and situated actions* (2nd ed.). Cambridge: Cambridge University Press.

Szpirko, J. (2000, July*). Can the categories of the Real, the Symbolic and the Imaginary be exported to other fields of knowledge than psychoanalysis?*

Presented at Les Etats Généraux de la Psychanalyse. Retrieved from http://www .etatsgeneraux-psychanalyse.net/TEXTE/Jean Szpirko/EGP. Site is no longer available. [Earlier version of article can be found in French as Szpirko, J. (1994). Les catégories RSI sont-elles exportables? *Les Carnets de Psychanalyse (Le Réel, la Réalité)* n° 5/6.]

Tajfel, H. (1970). Experiments in intergroup discrimination. *Scientific American*, 223, 96–102.

Tajfel, H. (1982). *Social identity and intergroup relations.* Cambridge: Cambridge University Press.

Tajfel, H., Billig, M., Bundy, R. P., & Flament, C. (1971). Social categorization and intergroup behaviour. *European Journal of Social Psychology*, 2, 149–178.

Tajfel, H., & Turner, J. C. (1979). An integrative theory of intergroup behavior. In W. G. Austin & S. Worchel (Eds.), *Psychology of intergroup relations* (pp. 33–47). Monterey, CA: Brooks/Cole.

Tashakkori, A., & Teddlie, C. (Eds). (2002). *Handbook of mixed methods in social & behavioral research.* Thousand Oaks, CA: Sage.

Tatum, B. (1997). *Why are all the black kids sitting together in the cafeteria?* New York: Basic Books.

Taylor, S. (2005). Self-narration as rehearsal: A discursive approach to the narrative formation of identity. *Narrative Inquiry*, 15, 45–50.

Thagard, P. (2008). *Hot thought.* Cambridge, MA: MIT Press.

Tissaw, M. (2010). A critical look at critical (neo)personalism: *Unitas multiplex* and the 'person' concept. *New Ideas in Psychology*, 28(2), 159–167.

Titchener, E. B. (1912). The schema of introspection. *American Journal of Psychology*, 23, 485–508.

Tobin, K., & Roth, W.-M. (2007). Identity in science: What for? Where to? How? In W.-M. Roth & K. Tobin (Eds.), *Science, learning, and identity: Sociocultural and cultural-historical perspectives* (pp. 339–345). Rotterdam: Sense Publishing.

Tolman, C. W. (1998). Sumus Ergo Sum: The ontology of self and how Descartes got it wrong. In W. E. Smythe (Ed.), *Toward a psychology of persons* (pp. 3–24). Mahwah, NJ: Erlbaum.

Tomasello, M. (1999). *The cultural origins of human cognition.* Cambridge, MA: Harvard University Press.

Tomasello, M. (2003). The key is social cognition. In D. Gentner & S. Goldin-Meadow (Eds.), *Language in mind: Advances in the study of language and thought* (pp. 47–57). Cambridge, MA: MIT Press.

Toulmin, S. (1972). *Human understanding: Vol. I. The collective use and evolution of concepts.* Princeton, NJ: Princeton University Press.

Toulmin, S. (1990). *Cosmopolis: The hidden agenda of modernity.* Chicago: University of Chicago Press.

Trafton, J. G., Trickett, S. B., & Mintz, F. E. (2005). Connecting internal and external representations: Spatial transformations of scientific visualizations. *Foundations of Science*, 10, 89–106.

Traweek, B. S. (1988). *Beamtimes and lifetimes: The world of high energy physicists.* Cambridge, MA: Harvard University Press.

Trickett, S. B., & Trafton, J. G. (2007). "What if . . . .": The use of conceptual simulations in scientific reasoning. *Cognitive Science*, 31, 843–876.

Trower, C. A., & Chait, R. P. (2002). Faculty diversity: Too little for too long. *Harvard Magazine, 104*(4), 33–98.

Turner, J. C. (1982). Toward a cognitive redefinition of the social group. In H. Tajfel (Ed.), *Social identity and intergroup relations* (pp. 15–40). Cambridge: Cambridge University Press.

Turner, J. C., Brown, R., & Tajfel, H. (1979). Social comparison and group interest in ingroup favoritism. *European Journal of Social Psychology, 9,* 187–204.

Turner, S. (1994). *The social theory of practices.* Chicago: University of Chicago Press.

Tversky, A., & Kahneman, D. (1974). Judgment under uncertainty: Heuristics and biases. *Science, 185,* 1124–1131.

Tweney, R. D. (1989). A framework for the cognitive psychology of science. In B. Gholson, W. R. Shadish Jr., R. A. Neimeyer, & A. C. Houts (Eds.), *Psychology of science: Contributions to metascience* (pp. 342–366). New York: Cambridge University Press.

Tweney, R. (1992). Stopping time: Faraday and the scientific creation of perceptual order. *Physis: Revista Internazionale di Storia Della Scienza, 29,* 149–164.

Tweney, R. D., Doherty, M. E., & Mynatt, C. R. (Eds.). (1981). *On scientific thinking.* New York: Columbia University Press.

Valian, V. (1999). *Why so slow? The advancement of women.* Cambridge, MA: MIT Press.

Varela, F., Thompson, E., & Rosch, E. (1991). *The embodied mind: Cognitive science and human experience.* Cambridge, MA: MIT Press.

Van Langenhove, L., & Harré, R. (1999). Positioning and the writing of science. In R. Harré & L. van Langenhove (Eds.), *Positioning theory* (pp. 102–115). Oxford: Blackwell.

Vanheule, S., & Verhaeghe, P. (2009). Identity through a psychoanalytic looking glass, *Theory & Psychology, 19*(3), 391–411.

Vosniadou, S., & Brewer, W. F. (1992). Mental models of the earth: A study of conceptual change in childhood. *Cognitive Psychology, 24,* 535–585.

Vygotsky, L. S. (1978). *Mind in society: The development of higher psychological processes* (M. Cole, Trans.). Cambridge, MA: Harvard University Press.

Watson, D. L. (1938). *Scientists are human.* London: Watts & Co.

Watson, J. (1913). Psychology as the behaviorist views it. *Psychological Review, 20,* 158–177

Watson, J. D. (1969). *The double helix.* New York: New American Library.

Wenger, E. (1998). *Communities of practice: Language, meaning, and identity.* Cambridge: Cambridge University Press.

Weick, K. E. (1995). *Sensemaking in organizations.* Thousand Oaks, CA: Sage.

Weigman, R. (1998). American anatomies: Theorizing race and gender (2nd Printing). Durham, NC: Duke University Press. Originally published 1995.

Wellman, H. (1990). *The child's theory of mind.* Cambridge, MA: MIT Press.

West, C., & Zimmerman, D. (1987). Doing gender. *Gender and Society, 1,* 125–151.

Wetherell, M. (1998). Positioning and interpretive reperatoires: Conversation analysis and post-structuralism in dialogue. *Discourse & Society, 9*(3), 387–412.

Wilkinson, S., & Kitzinger, C. (2003). Constructing identities: A feminist conversation analyst approach to positioning in action. In R. Harré & F. M. Moghaddam (Eds.), *The self and others* (pp. 157–180). Westport, CT: Praeger.

Wilson, R. (2004). *Boundaries of the mind: The individual in the fragile sciences.* Cambridge: Cambridge University Press.

Winner, E. (2000). The origins and ends of giftedness. *American Psychologist, 55*(1), 159–169.

Wittgenstein, L. (1953). *Philosophical investigations* (G. E. M. Anscome, Trans.). Oxford: Blackwell.

Wollheim, R. (1999). *On the emotions* (The Ernst Cassirer Lectures, 1991). New Haven: Yale University Press.

Xie, Y., & Shauman, K. A. (2003). *Women in science: Career processes and outcomes.* Cambridge, MA: Harvard University Press.

Young-Bruehl, E. (1996). *Anatomy of prejudices.* Cambridge, MA: Harvard University Press.

Zhang, J. (1997). The nature of external representations in problem solving. *Cognitive Science, 21*(2), 179–217.

Zuckerman, H., Cole, J. R., & Bruer, J. T. (Eds.). (1991). *The outer circle: Women in the scientific community.* New York: Norton.

INDEX

achievement in science, 12
act theory (Brentano), 19
"acting," 23. *See also under* unit(s) of analysis
action/activity, 18–20
    act(ion) vs. activity, 20
    activity vs. practice, 25–27
    as expressing coordination, overcoming
        dualisms, 21–24
    three-level model of activity, 20
actions, defined, 23
activity theory, 20–22
actor network theory, 83n2, 154
acts, defined, 23
adjustment and "adjusted" persons, 27–28
Adler, Alfred, 127
"affective revolution," 97
affectively toned metaphorical expressions,
        108. *See also* expressions, figurative and
        metaphorical
agency, 22–24, 218
    attributions of, 83n2, 147–50, 154
    Edward Pols on, 23
    material, 151–55
    right of, 144
    terminology, 197
"agentive," 197
agentive environment, 234
agentive intent, 154
agentive learning, 203, 235–36
agentive learning environments, 102, 107, 197,
        234
agentive unconscious, 235
agents, 197
    human and nonhuman, 197
Ajzen, Icek, 224
Alasuutari, Pertti, 37–38

Alcoff, Louise, 126–27, 170
Anderson, Michael L., 19
animat model-system, partial, 76, 77
Anscombe, Elizabeth, 29
anthropomorphism, 150, 208
    cognitive partnering and, 148
anthropomorphizing
    attributions of agency and, 82
    of cells, 203
    defined, 147
anthropomorphizing expressions, 102, 148, 155
    "happy" cells, 112–19, 147
apprenticeship enactments, 199–200
Aristotle, 20
artifact model, 63, 66, 68, 84–88
artifact model-system, 79
artifacts, 46
    cognitive, 54, 90
    cultural, 54, 90
    laboratory, 32, 101, 112, 117, 118, 196–98, 200
        positioning and, 146–55
artificial intelligence, 18, 19, 95
audit, external, 49

Baldwin, James, 169n10
Bennett, Max R., 244
Bickhard, Mark, 24
"biggest-picture goal," 70
biological engineers, 138
biomedical engineering (BME), 15
    as interdiscipline, 32–33
biomedical engineering (BME) laboratories,
        16, 32–33, 157
    emotion and motivation in, 101–2
biomedical engineers, 130
black identity, 170. *See also* racial identity

Bowlby, John, 99
Brentano, Franz, 19
Bridgman, Percy W., 24, 239

categorization, 221
category mistakes, 244
cell agency, 148, 154. *See also* agency
cell-culture line of research, 58
cell cultures, 57–58, 60, 61, 69, 89, 152
    solving the problem of contamination of,
        60
cell culturing, 56
    learning, 199, 201
cell-culturing techniques, 81, 201–2
cell death, 152–54, 156
cells. *See also* learning path
    anthropomorphizing of, 203
    cooperative relationship with, 148–51
    "happy," 112–19, 147
    progenitor, 63–66, 68, 69
child as scientist, 229n3
coding and analysis, 46–48
    grounded coding, 46–48
    rigor and plausibility, 49
cognition. *See also specific topics*
    as act and activity, 25
    emotional and embodied, 97–98
    "environmental perspectives" on, 53
    scientific (*see* scientific thinking)
cognitive apprenticeship, 81
cognitive artifacts. *See* artifacts, cognitive
cognitive-cultural systems, 83, 88, 91, 238
    distributed, 56, 79
cognitive emotions, 97
cognitive-historical analysis, 49–50
cognitive partnering, 79–83, 90, 118, 148–49
    anthropomorphism and, 148
    coupling and, 84
    defined, 79
    as form of positioning, 148–49
    Nersessian on the concept of, 15
    person-to-artifact, 79–80, 82–83
    person-to-person, 79–81
cognitive practices, 15
cognitive science, 7, 19
collaborative coding, 49
collective self-concept. *See* social identity
        theory
communities
    discursive, 129–30
    identities and, 120–21, 126
    innovation, 54, 102

compromise formations, acts as, 159
conscientiousness, 12
construct devices, 61, 62, 65, 68. *See also* devices
construct model, 61, 62. *See also* devices,
        construct
"construct within a construct," 80
constructive alternativism, principle of, 238
constructs and constructing. *See* personal
        construct theory
construing as holistic activity, 236–38
"cool," meanings and connotations of the
        word, 111
"cool" science, 110–12
cooperative relationship with cells, 148–51
coupling, 84, 87–88
Craik, Kenneth, 86
creative environments, laboratories as, 33
creativity, 12
Crick, Francis, 99–100
cultural artifacts. *See* artifacts, cultural
cultural rackets, 90
culture, 237. *See also* scientific thinking

data collection and analysis. *See also* coding
        and analysis
    issues posed, 36–42
    methods of analysis, 49–51
data collection procedures
    ethnographic observation, 43–44
    gender and race enactment data collection,
        44–45
    transcription and notation, 45
    triangulation of data, 45
De Mey, Marc, 8
de Sousa, Ronald, 93, 95, 96
Denzin, Norman, 38, 39
depressive realism, 95
desire, 10. *See also* emotion
detachment, ideal of scientific, 227, 228
developmental theory and identity, 123, 124
devices, 55–60, 107. *See also* cognitive
        partnering; models, interlocking
    construct, 61, 62, 65, 68
    dish, 35–36, 70–76, 78–79 (*see also* dish
        model-system)
    flow channel, 59
    flow loop, 57–68, 89
    as "hubs," 88, 90
    interactive practices developing around, 56
    physical models as, 55, 56
    simulation, 60, 82, 88, 117, 196
    "wet," 60

Dewey, John, 218, 240, 245
  analysis of technology, 20
  attack on the "spectator view" of knowledge, 19
  functionalism and, 28
  on interaction and situation, 20
  on knowing, 53, 231
  on pedagogy, 52
  Peter Godfrey-Smith on, 21
  pragmatism, 27, 28
  on problem solving, 52–53, 231
  on reflex arc, 21
  on representationalism, 25
  on scientists, 246
  on thinking, 25, 233
  transactional theory of emotion, 116
  on units of analysis, 21
dialectical engagement, 234–35
discourse(s)
  science, 165
  of science proper vs. improper, 175n11
discovery, scientific, 240
discursive analysis, 39, 46
discursive communities, 129–30
discursive events/productions, 40, 41, 46
discursive positioning, 138
discursive practices, 39, 155, 156, 165
discursive problem-space, interdisciplinary research laboratory as, 196–98
discursive strategies, 39–41, 126, 128, 130, 156, 235
dish devices, 35–36, 70–76, 78–79
dish model-system, 70, 72, 82–83
distributed cognition, 25, 31–32, 40, 69, 232
distributed cognitive-cultural systems, 56, 79
distributed cognitive system, 83
distributed model-based cognition, 83–88
distributed model-based problem solving, 64, 66
distributed model-based reasoning, 88, 90
distributed problem solving, 64, 66
  enacting, 79
  cognitive partnering, 79–83
  interlocking models, 88–90 (*see also* models, interlocking)
distributing knowledge, 33
Dreyfus, Hubert L., 98
Dreyfus, Stuart E., 98
Dunbar, Kevin, 41

education, 20, 52
Eiduson, Bernice, 11

Emirbayer, Mustafa, 116
emotion, 92, 119. *See also* expressions; integration problem
  cognition and, 97–98
  as indication of intimate participation, 117
  knowledge and, 9
  and motivation
    in biomedical engineering laboratories, 101–2
    connections between, 105–8
    overt expressions of, 102–8
  nature of, 93
  in psychology, 93–94
  rationality and, 94–96
  science and, 99–101
  sense-making and, 4, 92, 100, 101, 106, 108, 110, 112
  as transactional, 115–18
emotion words, 99–101
emotional cognition, 97
emotional neutrality. *See* neutrality
endothelial progenitor cells (EPCs), 66, 68, 69
Engeström, Yrjö, 20, 25
"environmental perspectives" on cognition, 53
environmentalist approaches in cognitive science, 10, 15
Epistemic Cultures (Knorr-Cetina), 2
Erikson, Erik H., 121
error, 244
  subjective/personal dimension as source of, 10–11
excitement, 102–4
expectations, similarity of, 237
experience, attributions of, 147–48
expressions, 119. *See also* anthropomorphizing expressions
  figurative and metaphorical, 108–12

Fanon, Franz, 165n3
Feist, Greg, 5, 11–14, 221, 243–44
feminist scholars, 9
Fischer, Agneta H., 116
flow channel devices, 59
flow loop, 58
  nature of, 133–34
flow loop devices, 57–68, 89
flow loop simulation, 89
Folkman, Susan, 115–16
Foucault, Michel, 39
Fransella, Fay, 231–32, 235, 236
Fridlund, Alan J., 116
frustration, 102, 104–7, 183

gender, 157–62
conceptual restraints regarding science and, 182
"double shift" in relation to theorizing about, 181–82
failure and, 183–85
"perfectionists" and floozies, 185–88
as performative, 181–83
personalizing research, 188–94
persons enacting, 179–81
gender competition, 214
gender discrimination, 160
gender-focused interview, 212–17
gender identity, 159. *See also* gender
generalized other, 141
Gestalt psychology, 240
Gibson, James J., 23
Gilbert, G. Nigel, 175n11, 184
Glaser, Barney G., 39
Godfrey-Smith, Peter, 21
Goffman, Erving, 127
Goldberg, Chad A., 116
Gorman, Michael, 13
grab-bag general psychology, problem of, 6, 16–17
Grene, Marjorie, 27
Griffiths, Paul E., 116
group feedback, 49

Hacker, Peter M.S., 244
happiness. *See* anthropomorphizing expressions, "happy" cells
Harré, Rom, 39, 93, 129, 130, 155, 224
Heiddeger, Martin, 39
Hesse, Mary, 247
heuristics, 224
Hewitt, Nancy M., 166–67
holistic framework, 236–38
Holton, Gerald, 186n20
Husserl, Edmund, 39
Hutchins, Ed, 33, 84
hybrot model-systems, 75

identities
communities and, 120–21, 126
contemporaneous integration of, 167
negotiation of, 29, 120, 122, 131, 157 (*see also* gender; race; social pact)
identity, 39, 47. *See also* race; social identity
cell death and, 154
concept of, 44, 120

conceptual history, 125–26
definitions, 125
developmental approach to, 123, 124
emotion and, 120, 219 (*see also* emotion; expressions)
as enactment, 158–59
George Herbert Mead and, 121, 141, 150
laboratories and, 208
learning and, 219
mixed, 138–39
nature of, 122
positioning and, 126–28, 131, 138 (*see also* positioning)
practice and, 126
psychology and, 120–26
"recovering," 158
identity formation(s), 50, 121, 123, 141, 197. *See also* agency
George Kelly on, 247
interpersonal encounters and shifts in, 245
identity production, language as medium of, 127
identity transformations, 198, 218
images, emotion-laden, 98
in vivo-in vitro divide, traversing the, 54
individualistic notions of knowledge generation, 9
innovation communities, 54, 102
innovation-seeking agenda of laboratories, 33
integration problem (science studies), 5–8
defined, 5
integrative efforts, 9–10
rational-social divides, 8–9
toward enhanced integration, 13–16
"intense absorption," 98
intent, agentive, 154
internal-external divide, 90
International Society for the Psychology of Science and Technology, 13
Isaacson, Walter, 13

James, William, 25, 28, 94, 242
Johnson-Bailey, Juanita, 169–70
joint attention, 171

Kant, Immanuel, 123n4
Keller, Evelyn Fox, 181
Kelly, George
person-as-scientist metaphor, 1, 2, 229–39, 247
Kinder, Donald R., 169

Kitcher, Philip, 4, 223, 225
Kitzinger, Celia, 164
Knorr Cetina, Karin, 2, 224
Kövecses, Zoltan, 108
Koyré, Alexandre, 53
Kruglanski, Arie W., 224
Kuhn, Thomas S., 22, 26, 226
Kurz-Milcke, Elke, 144

laboratory artifacts. *See* artifacts, laboratory
laboratory(ies), 31–34. *See also* cognitive-cultural systems
    as evolving systems, 33
    as innovation communities, 54
    innovation-seeking agenda, 33
    interdisciplinary research, 196–98
    Lab A, 34–35
    Lab D, 35–36
    nature of, 1–2
    objectives, 54
    as "problem-space," 32, 196–98
    social organization, 182–83
    storyline, 130
Lacan, Jacques, 44, 50, 51
Lamiell, James, 17
Lampkin, Richard, 222–23
Lashley, Karl, 144–45
"latching," 140
Latour, Bruno, 83n2, 129, 146, 154
Lave, Jean, 5–6, 24, 199–200
Lazarus, Richard S., 115–16
learning, 195–96. *See also* agentive learning
    by doing, 201–3
learning path, researchers' sense of their, 198
    case studies, 198–211
    defined, 195
Lee, Patrick, 99
legend, 223
legitimate peripheral participation, 199–200
Leont'ev, Aleksei N., 20, 21
Lewin, Kurt, 23
Liberman, Sofia, 13
linguistic analysis, positioning analysis as, 155
Longino, Helen, 8–9, 248

Mahoney, Michael J.
    on emotion and subjectivity, 99, 241–44, 246
    on psychology of science, 14
    on scientists, 4, 223, 225, 241
Malone, Kareen R., 44–45, 50

Manstead, Antony S. R., 116
Manuel, Frank E., 99
Margolis, Joseph, 222
Martin, Jack, 23
Marx, Karl, 20
Maslow, Abraham H., 228–29, 239, 243, 245
Mcguire, James E., 245
Mead, George H., 20, 23, 121, 141, 149, 150, 231
mental model, 90
    meanings of the term, 85
mental simulation, 85
mereological fallacy, 244
methodology, 7. *See also* data collection and analysis; qualitative methods in psychology; *specific topics*
Meyerson, Emil, 53
mezzo-level neuroscience, 70
Miettinen, Reijo, 20, 25
Mills, Charles, 169n10
mind, theory of, 237n4
mind-world relations, 18
Mitroff, Ian
    on emotion, subjectivity, and personhood of scientists, 99, 241–44, 246
    on "storybook" view of science, 4, 222, 223, 241, 242
model-based cognition, 79
    distributed, 83–88
"model-based cognizing," 48
model-based reasoning, 90, 91
    distributed, 88, 90
    vs. model-based understanding, 48
    simulative, 55
model-based simulation, 54, 79
model-based understanding, 48, 206
model-system configuration, vascular-construct, 63–69
model-systems, 55–56. *See also* problem solving
models. *See also under* problem solving
    interlocking, 65, 75, 86, 86–91
Moghaddam, Fathali M., 121, 130
motivation. *See* emotion, and motivation
Mulkay, Michael J., 175n11, 184

Nelson, Donna J., 166
Nersessian, Nancy J., 50, 144, 238
    on cognitive partnering, 15
    laboratories and, 4, 15, 16, 33, 36
    on mental models, 36n6, 85
    on rational-social dichotomy, 8–9
    scientific contributions, 220

neutrality, 228, 242. *See also* value-free and value-neutral science
Newstetter, Wendy C., 4, 144
  laboratories and, 4, 15, 16, 36
  scientific contributions, 220
norms, 10

objectivity, 227
  vs. concern, 228
  nature of, 9
Olson, Kristen, 169
Ortner, Sherry, 42

Papadopoulos, Dimitris, 22
paradigms of psychological thought, 22
Parkinson, Brian, 116
Parrott, W. Gerrod, 93
participation
  emotion as indication of intimate, 117
  legitimate peripheral, 199–200
peer-to-peer partnering, 80
people- vs. thing-orientation, 11–12
peripheral participation, legitimate, 199–200
person-as-scientist metaphor, 1, 2, 29, 184, 229n3, 230, 232, 238. *See also* personal construct theory
  Black's interactionist view of metaphor and, 247
  implications of, 238, 247
person-to-artifact relations, 147. *See also under* cognitive partnering
personal construct theory (PCT), 229–38
  transformative power, 235
Personal Knowledge, Science, Faith and Society (Polanyi), 226–27
personality and science, 11–13
"personality equation," 10
"personality problem," 10–11
personhood and science, 239–46
philosophy and social sciences, 8
Piaget, Jean, 53
plasticity experimentation, 105–6
Polanyi, Michael, 99, 226–28, 239, 240, 243, 246
Pols, Edward, 23
Popper, Karl, 53
positioning. *See also under* social pact
  cognitive partnering as form of, 148–49
  in the laboratory, 129–31
  laboratory objects and artifacts and, 146–55
  professional and academic identity and, 131–40
  as warrant and justification, 140–46

positioning theory, 126–29, 164
positions, features of, 128
practice, 1, 24–27. *See also* science practice
pragmatism, 27–28
problem solving, 6, 52–54, 90. *See also* distributed problem solving; integration problem; *specific topics*
  carried out in conjunction with environment, 31
  Dewey on, 52–53, 231
  integrative scheme for analyzing, 54–57
  in neural engineering laboratory, 69–70 (*see also* devices, dish)
    animat model-system configuration, 75–79
  in tissue engineering laboratory, 57
    the construct, 60–62
    flow loop, 57–60, 62
    vascular-construct model-system configuration, 63–69
problem-solving persons, 53
progenitor cells, 63–65
  endothelial, 66, 68, 69
progress, 222
Protevi, John, 19
prototype, 5, 221
  as ideal, 221–25
  as typical, 221, 225–26
psychoanalytic analysis, 50–51
psychology. *See also specific topics*
  qualitative methods in, 36–42
  relations to science, 3
  as science vs. science as psychology, 3–5 (*see also* science, as psychology)
    problems in studying, 5–7
  scientific, 3
  scope of the term, 16
  "second," 125
  as theoretical pursuit, 6
Psychology of Personal Constructs, The (Kelly), 229, 237
"psychology of science," 239
  field of, 7, 14, 220
  as research focus, 220, 221
  use of the term, 13, 14
Psychology of Science, The (Maslow), 228, 239
Psychology of Science and the Origins of the Creative Mind (Feist), 5

qualitative data and analysis, 36–42
qualitative methods in psychology, 36–42

race, 157–62, 185. *See also* social pact
  persons enacting, 162–65
  science and, 167–72
racial diversity in science, problem of, 165–67
racial identity, 163–64, 170. *See also* race
racism, symbolic, 169
rational-social dichotomy, 8–9
rationality, 5
reflexivity and personal construct theory, 235
Reichenbach, Hans, 100
representational view, 18
representationalism, 25
  alternatives to, 19
representationalist accounts of mind-world relations, 19
representations, processes, and S-R connections vs. acts, actions, activities, and practices, 18–20
"representations over computations" doctrine, 18
representing vs. representation, 85
researcher-artifact-animal models, system of, 69
researcher model, 63, 66, 84, 86–88
robots, "real," 103
Rogers, Diana C., 166
roles. *See* social role
Rosch, Eleanor, 226
Rosser, Sue V., 180
Rouse, Joseph, 25–27, 41
Rowlands, Mark, 15
Russell, James A., 116
Rychlak, Joseph, 234

Scarantino, Andrea, 116
Schaffer, Simon, 10–11
Schatzki, Theodore R., 24
science
  vs. achievement in science, 12
  as activity and practice, 1, 24–27
    qualifiers, 27–28
  "cool," 110–12
  culture of, 162n1
  engaged, 226–29
  as human practice, 220
  ideals of, 4, 224, 227
  nature of, 225
  philosophy of, 8
  Polanyi's "conceptual reform" of, 227
  as prototypical, 229
  as prototypically "human," 221
  as psychology, 229, 246–48
  and psychology of science, 220–21
  scientists' disillusionment with, 182
  Skinner on methods of, 224
  types of, 2–3
science practice, 1, 4, 28, 106, 111, 112, 170, 171
  agency in, 93
  analysis of, 221
  approach to learning in, 196
  cultural negotiation and, 157
  dimensions of, 2, 41, 238
  diversity and, 165
  gender and, 161, 181
  gender differences in, 179, 180
  gender enactment in, 188
  identity and, 168
  material grounding of, 10
  normativity and, 26
  person-centered account of, 245
  positioning and, 129, 158
  social pact of, 158
  "something else" in, 10, 11, 13, 238
science practices, researchers' accounts of their, 219
science proper vs. improper, discourses of, 175n11
"science wars," 8
scientific attitude, historical development of the, 243–44
"Scientific Attitudes" (Lampkin), 223
scientific change as problem-solving process, 53
scientific detachment, ideal of, 227, 228
scientific method, 7, 224
scientific thinking, socio-cultural and rational-cognitive accounts of, 5–9. *See also* Dewey, on thinking; integration problem
  integrative efforts, 9–10
  omissions and oversights, 10–11
scientism, 3, 12, 222, 238
Scientist as Subject (Mahoney), 242–43
scientist(s), 2, 223. *See also* person-as-scientist metaphor
  acting person as (*see* unit(s) of analysis)
  as actor, not reactor, 232–34
  acts holistically, 236–38
  as the chief problem of science, 239
  compared with the "man in the street," 246
  construes at different levels of awareness, 235
  cultural depictions of, 225
  engaged dialectically, 234–35

scientist(s) (*cont.*)
　formation, 182
　George Kelly on, 1, 230
　as inveterate innovator, 238
　as metaphor for person, 1
　Michael Mahoney on, 4, 223, 225, 241
　personality traits, 11–13
　as reflexive, 235–36
Scientists: Their Psychological World (Eiduson),
　　11
Scientists Are Human (Watson), 239–40
Sears, David O., 169
self-categorization, 122. *See also* identity
self-concept, 122, 123. *See also* identity
self sense-making, 122
Selz, Otto, 53
sense-making, 27, 29, 39, 42, 47, 197, 208, 221.
　　See also learning path; positioning
　brain, environment, and, 31
　emotion and, 4, 92, 100, 101, 106, 108, 110, 112
　examples of, 106, 142, 201, 235
　George Kelly, personal construct theory,
　　and, 230–32, 235, 237, 247
　imaginative dimension, 23
　integrated account of social identity and,
　　247
　mental models and, 85
　required at every step, 248
　self sense-making, 122
sense-making orientation, 230, 232
Seymour, Elaine, 166–67
similarity of expectations, 237
Simon, Herbert, 53
Simonton, Dean Keith, 12
simulation
　flow loop, 89
　mental, 85
　model-based, 54, 79
simulation devices, 60, 82, 88, 117, 196
simulation models, 107
simulative model-based reasoning, 55
simulative modeling, capacity for, 85
situated activity, 19
situated cognition, 25, 27, 31, 40, 49
situated learning, 217
situated theory of emotion, 116
skill acquisition, cognitive basis of, 98
Smith, Jonathan A., 37
Smythe, William, 24
social identity, 121
　vs. personal identity, 121, 123

social identity theory, 121–23, 163
social meanings, denied, 168
social pact, 175n11
　positioning, marginalia, and the, 172–78
social role, 127–28
sociality, 237
Sonnert, Gerhard, 186n20
"speaking rights," 169n10
"spectator view" of knowledge, 19
spike detector, 74
Spradley, James P., 44
Stern, William, 17–18, 23
Sternberg, Robert, 16–17
Storer, Norman W., 226
"storybook" view of science, 4, 222, 223, 241,
　　242
Strauss, Anselm L., 39
subjectivity/subjective dimension, 227, 241, 244
　emotion and, 99, 241–44, 246
　influence of, 10–11
Sugarman, Jeff, 23

Tajfel, Henri, 122–23
Thagard, Paul, 92, 97, 99–101
Thompson, Janice, 23
Titchener, Edward B., 11
Tolman, Charles W., 23
Tomasello, Michael, 171
trace identity effects, 167
transactional theories of emotion, 115–18
Tuchanska, Barbara, 245
Turner, John C., 121

uncertainty, 224, 239
unitas multiplex, 17–18
unit(s) of analysis, 15, 17, 18, 22
　acting and activity as, 20, 21
　acting person (as scientist) as, 23, 24, 27, 29,
　　39, 120, 195, 246
　activity theory and, 21–22
　of cognitive practice, 32
　vs. constructs, 231
　integrated nature of, 21
　in personal construct theory, 231
　positioning theory and integrated, 155
　practice as, 24
　in the study of emotions, 116

value-free and value-neutral science, 4
values, 10
Van Langenhove, Luk, 129, 155

vascular-construct model-system, partial, 66, 67, 68
vascular-construct model-system configuration, 63–69
Vygotsky, Lev Semyonovich, 20, 21, 22, 23, 127

Watson, David L., 239–40, 243, 244, 246
Watson, James D., 99–100
Weick, Karl E., 29

Wenger, Etienne, 120, 125–26
Who Succeeds in Science (Sonnert and Holton), 186n20
Wilkinson, Sue, 164
Wittgenstein, Ludwig, 20, 39
women. *See also* gender
    in academe and science research, 182
working memory, 85, 87
Würtzburg school of psychology, 53